20 070 709 60
WITHDRAWN

AF615882

Real-time Systems' Quality of Service

Roman Gumzej • Wolfgang A. Halang

Real-time Systems' Quality of Service

Introducing Quality of Service Considerations in the Life-cycle of Real-time Systems

Roman Gumzej, Dr.
University of Maribor
Faculty of Electrical Engineering
and Computer Science
Smetanova ulica 17
SI-2000 Maribor
Slovenia
roman.gumzej@uni-mb.si

Wolfgang A. Halang, Prof. Dr. Dr.
Fernuniversität in Hagen
Lehrgebiet Informationstechnik
58084 Hagen
Germany
wolfgang.halang@fernuni-hagen.de

ISBN 978-1-84882-847-6 e-ISBN 978-1-84882-848-3
DOI 10.1007/978-1-84882-848-3
Springer London Dordrecht Heidelberg New York

British Library Cataloguing in Publication Data
A catalogue record for this book is available from the British Library

Library of Congress Control Number: 2009936773

Cover design: eStudioCalamar, Figueres/Berlin

Printed on acid-free paper

Springer is part of Springer Science+Business Media (www.springer.com)

"Whatever ye sow, so shall ye reap."
(The Bible, Galatians 6:7–8)
So we plant an idea in good will, and hope to grow a tree of wisdom.

Foreword

The discussion about the "quality" of software and software-based systems is certainly as old as the subjects themselves. However, this discussion has not yet been very fruitful, mainly because different people attribute different meanings to the notion of quality. For some it consists in the novelty of the algorithms used, for others in the use of formal verification methods, or in that of a particular programming language. As far as the majority of people is concerned, it seems that they would judge, however, the "quality" of a system on the basis of the number of "features" it offers to the user. Unfortunately, this attitude has led to gigantic, clumsy, and nearly unmanageable software systems that are (necessarily) full of bugs. One does not need to name them, because on one hand everybody knows them, and on the other hand their vendors have a reputation for launching lawsuits against critics. At least, not much harm is done by such systems as long as they stay off-line and just anger their users by means of computer breakdowns, loss of (sometimes important) data or nasty hacker attacks. The market just gets what it demands and has to live with it.

The situation has to be judged differently, however, as soon as real-time systems come into play. In the first moment, an average user might react by saying: "Well, as I am neither supervising a power station nor piloting an airplane, why should I bother?" Unfortunately, this is not true. Today, an average car contains more software than an airplane did 30 years ago. Most of this software has the character of real-time software. Of course, parts of this software are not critical at all like, e.g. that of navigation systems. The worst it can do to you is not finding your own home if you happen to live in a small side street (as has happened to the author of this Foreword). Others may represent a safety hazard like, e.g. those that support the use of mobile telephones during driving. But there are very critical real-time systems in a car too like, e.g. the motor management. They are extremely useful, because they have greatly reduced fuel consumption and, additionally, enabled car engines to automatically adapt to low-quality fuel. However, they are critical as well, because, if they do not work properly, one may be in serious trouble. Modern car engines are often not able to run without electronic control anymore. This is in contrast to the situation in the beginning of the computer control era, when critical real-time systems were usually backed up by some analogue control circuits. Of course, this did not provide for the full functionality of the digital systems, but at least kept the plants in some kind of running order.

Unfortunately, it appears as if the market does not want to pay for this quality anymore. An example shall illustrate this. Some years ago, the author attended a colloquium on car electronics. One of the industrial speakers explained the functional principles of motor management systems very well. After his presentation a member of the audience asked what would happen in the case of a computer failure. He answered that the program would be able to detect this and shut the engine down. The audience became somewhat unquiet and somebody else asked what a driver should do if this were to happen during an attempt to overtake another car at high speed. The answer was that "this would certainly represent a problem, but accidents have always happened". As the audience was audibly dissatisfied with this answer, the speaker added that his customers were by no means willing to pay the extra cost for some redundancy.

Such observations confirm that this book is a necessity and fulfils an extremely important purpose. It may help to trigger a necessary change in the perception of software quality. Quality of software is no longer a diffuse entity with too many aspects to be dealt with successfully. Readers are directed towards the concentration on a particular subarea, i.e. Quality of Service (QoS), the subject of this book. This does not mean that other aspects may be neglected, but as a beginning something concrete can be identified and handled in a rational way. For practical applications it also helps a great deal that applicable standards and benchmarks are described in sufficient detail, and that the material is partially presented in the form of checklists. In this context it does not matter whether the author of this Foreword completely agrees with the authors of the book in every aspect or not. Their basic approach is so important that a more or less "academic" discussion about one detail or another appears to be counterproductive. One particular detail shall even be particularly mentioned as a real progress in the discussion about software technology: for real-time systems software must always be designed and evaluated together with the underlying hardware — as, e.g. indicated in Sects. 6.1.4 and 7.1.1.

Thus, I would like to express my hope that this book will be well received by the technical community and, thus, be able to contribute to a necessary step towards real-time systems of higher quality, reliability and safety.

Munich, May 2009 *Peter F. Elzer*

Preface

Quality of Service (QoS) is gaining importance in the global market and, as such, is affecting all available products. While, at first, merely the quality of material was important, with "high-tech" products QoS encompasses an integral measure of efficiency, economy and ease of use. Since production processes largely affect the quality of products, interest now focuses on computerised control systems for which a number of quality criteria must be observed during their real-time operation. Hence, we would like to measure the QoS of real-time computing systems in order to determine their suitability for certain control applications and the systems being controlled. We would also like to compare such computing systems with one another. Finally, we would like to certify systems fulfilling QoS requirements with corresponding quality marks.

So far, QoS measures for real-time systems have not received much attention, and the quality of real-time systems has been evaluated differently in different application areas mainly on an ad hoc basis. An assessment of some existing QoS measures for real-time systems will reveal that they are predominantly inappropriate. Taking a closer look at the fundamental issues of real-time systems, it will become clear that for (hard) real-time systems qualitative characteristics are much more important than quantitative metrics. The major qualitative performance measures are compiled in this book. Systems fulfilling the QoS criteria may finally be compared on the basis of costs.

In this book, the state of the art in QoS evaluation of real-time systems shall be reviewed. It will turn out that for their proper assessment a more thorough consideration of the nature of real-time systems is necessary. This will finally lead to a number of QoS criteria. Surprising to those who used to confuse real-time with fast computing, most of these criteria are not concerned with system performance, but are rather qualitatively comparative and exclusive, i.e. a system may either fulfil them or not. Based on these criteria the critical points in the design and development process of real-time systems, where these criteria should be applied or checked and how, are determined. As a conclusion, software development and certification standards are assessed, and a proposition with guidelines on how the presented criteria should be applied in the design, development and certification process of real-time computing systems is laid out.

Maribor and Hagen,
June 2009

Roman Gumzej
Wolfgang A. Halang

Acknowledgements

I would like to express my gratitude to the people who accompanied and supported me on my productive path, and without whose encouragement I would not have made it this far. Especially, I would like to thank my family and my co-author, mentor and friend Prof. Wolfgang A. Halang, who saw beyond my horizon and gave me meaningful directions when I could not see the lighthouse anymore. Thank you all!

Roman

Contents

Acronyms

3GPP	3rd Generation Partnership Project
3GPP2	3rd Generation Partnership Project 2
ACID	Atomicity, Consistency, Isolation, Durability
AES	Advanced Encryption Standard
ATM	Air Traffic Management
(B)IA	(Business) Impact Analysis
BIT	Built-In Test
BIST	Built-In Self-Test
BSR	Boundary Scan Register
CAN	Controller Area Network
CMM	Capability Maturity Model
CORBA	Common Object Requesting Broker Architecture
CPU	Central Processing Unit
CSR	Communicating Shared Resources
D/T/M	Design/Testing/Maintenance
DECT	Digital Enhanced Cordless Telecommunication
DES	Data Encryption Standard
DIN	Deutsches Institut für Normung
DMR	Dual Modular Redundant
DSA	Digital Signature Algorithm
DTI	Department of Trade and Industry
DUT	Device Under Test
E/E/PE	Electrical/Electronic/Programmable Electronic
EDF	Earliest Deadline First
EDGE	Enhanced Data rates for GSM Evolution
ETSI	European Telecommunications Standards Institute
EUC	Equipment Under Control
FDD	Frequency Division Duplex
FMS	Fieldbus Message Specification
FOD	Failure On Demand
FORTRAN	FORmula TRANslator

FT	Fault Tolerance
GCM	Generic Component Model
GPRS	General Packet Radio Service
GRM	Generic Resource Modelling
GQAM	Generic Quantitative Analysis Modelling
GSM	Global System for Mobile communication
HRM	Hardware Resource Modelling
I/O	Input/Output
IDEA	International Data Encryption Algorithm
IEC	International Electrotechnical Commission
IEEE	Institute of Electrical and Electronics Engineers
ISO	International Organisation for Standardisation
ITU	International Telecommunication Union
JTAG	Joint Test Action Group
LLC	Logical Link Control
LLF	Least Laxity First
LFSR	Linear Feedback Shift Register
MAC	Media Access Control
MARTE	Modelling and Analysis of Real-Time Embedded systems
MD2	Message Digest 2
MD4	Message Digest 4
MD5	Message Digest 5
MDA	Model Driven Architecture
MTBF	Mean Time Between Failures
MTTF	Mean Time To Failure
MTTR	Mean Time To Repair
NFP	Non-Functional Properties
NPL	Network Perimeter Layer
OCL	Object Constraint Language
OLAP	On-Line Analytical Processing
OMA	Open Mobile Alliance
PAM	Performance Analysis Modelling
PAMR	Public Access Mobile Radio
PEARL	Process and Experiment Automation Realtime Language
PKI	Public Key Infrastructure
PMR	Professional Mobile Radio
POSIX	Portable Operating System Interface
PSA	Parallel Signal Analyser
QoS	Quality of Service
QT	Queuing Theory
RAID	Redundant Array of Independent Disks
RC2	Ron's Code 2 or Rivest Cipher 2
RC4	Ron's Code 4 or Rivest Cipher 4
RFID	Radio Frequency IDentification
RM	Rate Monotonic

RMA	Rate Monotonic Analysis
ROOM	Real-time Object-Oriented Modelling
RSA	Real-Time CORBA Schedulability Analysis
RSM	Repetitive Structure Modelling
RT	Real-Time
RT_STAP	Real-Time Space-Time Adaptive Processing
RTEMoCC	RTE Model of Computation and Communication
RTOS	Real-Time Operating System
S&A	Sensing and Actuation
SAM	Schedulability Analysis Modelling
SCADA	Supervisory Control and Data Acquisition
SHA	Secure Hash Algorithm
SIL	Safety Integrity Level
SSL	Secure Sockets Layer
TAP	Test Access Port
TDD	Time Division Duplex
TETRA	TErrestrial Trunked RAdio
TLS	Transport Layer Security
TLS-PSK	TLS Pre-Shared Key cypher suits
TLS-SRP	TLS Secure Remote Password
TMR	Triple-Modular Redundant
TPC-B	Transaction Processing Performance Council – Type B
TPC-C	Transaction Processing Performance Council – Type C
TVL	Tagged Value Language
UAV	Unmanned Aerial Vehicle
UML	Unified Modelling Language
UTRA	Universal Terrestrial Radio Access
VPN	Virtual Private Network
VSL	Value Specification Language
WAP	Wireless Application Protocol
WCEP	Worst Case Execution Path
WCET	Worst Case Execution Time
WLAN	Wireles Local Area Networks
WRAN	Wireless Regional Area Networks

1

Introduction

With our increasing dependence on computerised control systems, our concern about performance and safety of these systems is growing. When deciding on the selection of data processing components and systems we have to be aware of their benefits and hazards for the given application and make corresponding choices. The decision criteria differ from application to application and concern more than just processing speed. They include the following:

- Dependability and reliability
- User friendliness (learnability, ergonomics, help function, etc.)
- Maintainability (clear structure, well defined interfaces, etc.)
- Flexibility (program changes, upgradability, parameter changes, etc.)
- Services (available services, available connections, etc.)
- Safety (robustness, correctness, security mechanisms, e.g. against intrusions, etc.)
- Data security (authorisation mechanisms and encryption)
- Performance (speed, response time, real-time behaviour, memory requirements, etc.)

Some of these properties can be analysed by benchmark or monitor programs. One would also like to measure the performance and safety of real-time computing systems and compare them with one another. The suggested approach is to build a special requirements profile for any application to perform the specific evaluation. Here, the mentioned evaluation criteria can be used as guidelines. Since one is usually confronted with several alternative solutions, one would like to compare the different solutions in order to make decisions. Based of the results of their evaluation, one would also like to certify systems that fulfil certain performance and safety requirements with corresponding quality marks.

1.1 "Performance" of Real-time Systems

How should the performance of a real-time computing system be defined? According to [72], apart from fast computation, real-time systems also require higher input/output bandwidth and faster response than general-purpose systems (see Fig. 1.1). However, is mere CPU speed really important? Is I/O speed sufficient? Would more memory improve data processing? As can be seen, it is not an easy matter to measure the performance of a computer system, let alone if the computer system is operating in the real-time mode, whose most important characteristic is that computing tasks must be initiated at and completed not later than given points of time.

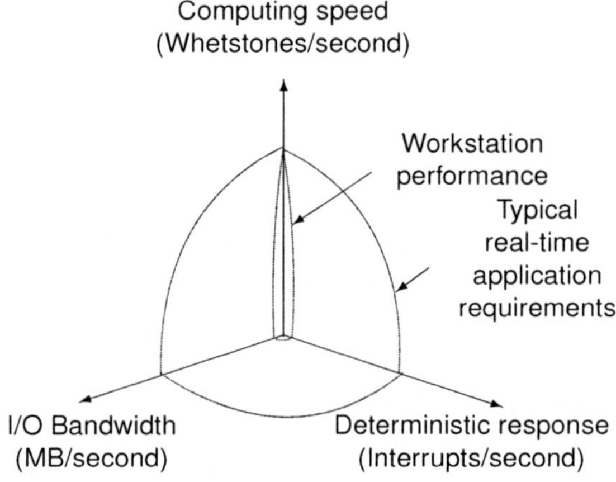

Fig. 1.1. Requirements space of real-time applications

As is shown in Fig. 1.2, a real-time system is defined as a special-purpose, concurrently working computing system that responds in a time-bounded way to specified external circumstances. These circumstances result from events occurring in the environment of the real-time computer system. Usually, a response has the form of sending actuation messages to the environment. The time restrictions are interpreted here as within and throughout specified time limits.

1.2 Requirements of Real-time Applications

Real-time applications impose many requirements on the different parts of real-time systems. The most fundamental requirement of real-time applications is the ability of the systems employed to respond to external events with very short and deterministic delays. Real-time applications usually run with critical deadlines. Therefore, a real-time operating system (RTOS) must have, among others, the following characteristics: a fully pre-emptable kernel, feasible task scheduling and graceful performance degradation in response to malfunctions. It should have deterministic and

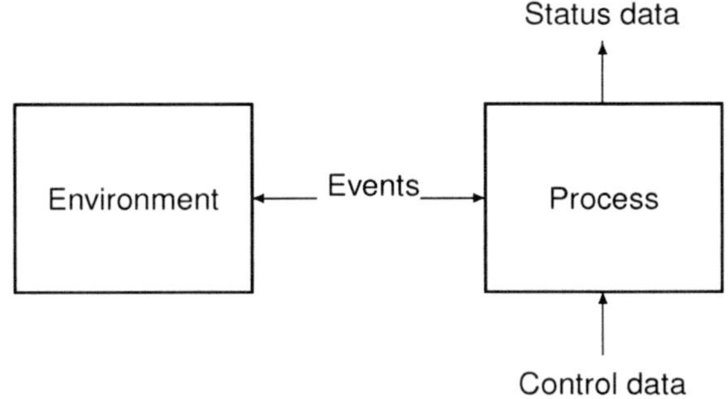

Fig. 1.2. Schematic definition of real-time computing systems

a priori known service times of system calls, which is particularly critical for synchronisation mechanisms. Furthermore, resource acquisition times are required to be bounded.

A number of requirements must be fulfilled by a real-time programming language. Among other things, it must provide constructs to communicate with the environment and with other tasks, possibly in a distributed computing system. It must provide access to a notion of time. To enhance reliability, it must support redundancy and diversity. Another important requirement for a programming language is that it must facilitate the generation of executable code with deterministic and predictable execution behaviour. For this purpose, certain features that may result in non-determinism must be avoided or renounced. For the most realistic program run-time estimations it is convenient that an analyser is built into the compiler, where internal program structures are known. An application structure should also support overload handling strategies (e.g. graceful degradation) by providing configurations of task sets with the necessary functionality, but with less demanding resource and/or timing constraints than the original configurations.

1.3 Real-time Systems Revisited

Since the term real time is, unfortunately, often misused, it is useful to mention here the precise definition of the real-time operating mode as given in the German industry standard DIN 44300 [14]:

> *The operating mode of a computing system in which the programs for the processing of data arriving from the outside are permanently ready, so that their results will be available within pre-determined periods of time; the arrival times of the data can be randomly distributed or be already a priori determined depending on the different applications.*

It is, therefore, the task of computers working in this operating mode to execute programs that are associated with external technical processes. As program processing must be temporally synchronised with events occurring in the external processes and must keep pace with them, real-time systems are also known as *reactive systems*. Furthermore, since real-time systems are always to be considered as being embedded in larger computing/control environments, they are correspondingly called *embedded systems* as well.

The explicit involvement of the dimension *time* distinguishes real-time operation from other forms of data processing. This is expressed by the following two fundamental user requirements, which real-time systems must fulfil even under extreme load conditions:

Timeliness: the ability to meet deadlines
Simultaneity of operation with the environments

As we shall see below, these requirements are supplemented by two further ones of equal importance:

Predictability: the ability to meet deadlines even under worst-case conditions
Dependability: the ability to ensure safe operation

When requested by external processes, data acquisition, evaluation and appropriate reactions must be performed on time. Not the mere processing speed is decisive for this, but the *timeliness* of reactions within pre-defined and *predictable* time bounds. Hence, it is characteristic for real-time systems that their functional correctness not only depends on the processing results, but also upon the instants when these results become available. Correct instants are determined by the environment, which cannot be forced to yield to a computer's processing speed, as is the case with batch and time-sharing systems.

External technical processes to be controlled usually consist of several more or less independent subprocesses operating in parallel. This implies the requirement for *simultaneous* processing of external process requests. To meet it, real-time systems must be essentially distributed, and must also provide parallel processing capabilities.

According to the types of the external processes, into which real-time systems may be embedded, there are environments with *hard* and with *soft* time constraints. As depicted in Fig. 1.3, they are distinguished by the consequences of violating the timeliness requirement. Whereas soft real-time environments are characterised by costs rising with increasing lateness of results, such *lateness* is not permitted under any circumstances in hard real-time environments, because late computer reactions are either useless or even dangerous. In other words, the costs of missing deadlines in hard real-time environments are — from the applications' point of view — (almost) infinitely high. Hard time constraints can be determined precisely, and typically result from the physical laws governing the technical processes being controlled [24]. On the other hand, as long as results arrive before their deadlines, they are temporally correct, and there is no value attributed to earlier arrivals.

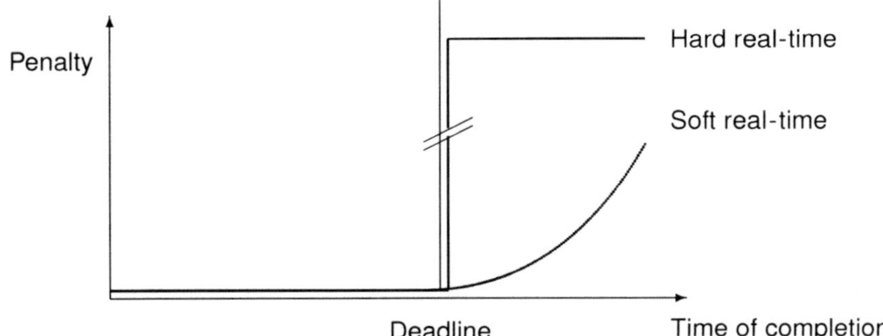

Fig. 1.3. Real-time environments with hard and soft timing constraints

1.4 Quality and Performance of Real-time Systems

The definition of the real-time operating mode has important consequences for the *dependability* of real-time systems, since the demanded permanent readiness of computers working in this mode can only be provided by fault-tolerant and — especially with respect to inadequate handling — robust systems. These dependability requirements hold for hardware and software alike. They are particularly important for those applications where computer malfunctions not only cause loss of data, but also endanger people and major investments. Naturally, expecting high dependability does not mean the unrealistic demand that systems never fail, since no technical system is absolutely reliable. Using appropriate measures, however, one must strive to make the violation of deadlines and corresponding damages quantifiable and as unlikely as possible. In doing so, the individual limitations of real-time systems must be recognised, and the risk of their utilisation in process control must be pondered carefully [24].

According to the definition of the real-time operating mode, data to be processed may arrive randomly distributed in time. This fact has led to the widespread conclusion that the behaviour of real-time systems cannot and may not be deterministic. This conclusion is, however, *based on a thinking error*. Indeed, an external technical process may be so complex that its behaviour appears to be random. The reactions to be carried out by the computer, however, must be precisely planned and fully predictable. This particularly holds for the case when several events occur simultaneously, leading to a situation of competition for service, and also includes transient overload and other error situations. Then, the user expects that the system will degrade its performance gracefully in a transparent and foreseeable way. Only fully deterministic system behaviour will ultimately enable the safety licensing of programmable electronic systems for safety-critical applications. Predictability of system behaviour is, as we have seen, of central importance for the real-time operating mode. It supplements the timeliness demand, as the latter can only be guaranteed if a system's execution behaviour is precisely predictable, both in time and with respect to external event reactions.

The definition of real-time operation as cited above implies that some prevailing misconceptions [71] about real-time systems need to be overcome: neither time-sharing nor just fast systems are necessarily real-time systems. It is important to stress that the evaluation criteria for embedded real-time systems are quite different from the ones used with regard to batch-processing or time-sharing computer systems. Commencing the processing of tasks in a timely fashion is much more significant than (average) speed. Thinking in probabilistic or statistical terms, which is common in computer science with respect to questions of performance evaluation, is as inappropriate in the real-time domain as the notion of fairness for the handling of competing requests or the minimisation of average reaction times as an optimality criterion of system design. Instead, worst cases, maximum run-times and maximum delays need to be considered, deadlocks must be prevented and deadlines must be observed under any circumstances. Systems may have no features that take arbitrarily long to operate. For the realisation of predictable and dependable real-time systems reasoning in static terms and recognising physical constraints is essential.

Maximum processor utilisation seems to be a major issue by the thinking criteria of computer science and is still the subject of many research papers in the domain of real-time scheduling. For embedded real-time systems, however, it is totally irrelevant whether processor utilisation is only suboptimal and throughput is rather low. What the user expects instead is dependable and predictable fulfilment of given requirements. Also, costs of a process control computer have to be seen in a larger context, viz., in the framework of the external technical process it controls. When a computer cannot guarantee reaction times, it may be unable to cope appropriately with exceptional and emergency situations arising in the process and may, thus, violate the latter's safety requirements. In comparison to the costs of a technical process and of possible damage that an overload or malfunction of a controlling computer may cause, *the price of a processor is usually negligibly low* and, in the light of the still constantly declining hardware costs, this price is of decreasing significance also in absolute figures.

An example from industrial practice is the one hour production stoppage of a medium-sized chemical facility due to a computer malfunction and the required clean-up incurring costs of some € 100,000. This is about the price of the entire computing system controlling the process itself. A processor board costs only a minor fraction of that amount, viz., some € 300, and the price of the proper processor is as insignificant as from few euros to few hundred euros. Lower processor utilisation is a — cheap — price to be paid for system and software simplicity being a prerequisite to achieve dependability and predictability. In the light of the fact that an experienced software engineer costs some € 100 for a line of tested and documented code produced, i.e. *the price of just one single program statement equals that of a CPU*, software simplicity and understandability have become the optimisation criteria of highest economical importance.

1.5 Preview and How to Use this Book

To start with, in the next chapter we shall review and assess the state of the art in performance evaluation of real-time systems. It will turn out that this state is rather inappropriate, necessitating a more thorough consideration of the nature of real-time systems, which will finally lead to a number of performance criteria. Surprisingly for those who used to confound real-time with fast computing, most of these criteria are qualitative or exclusive, i.e. a system may either fulfil them or not.

Then, we shall classify the pertaining "quality criteria for real-time systems". Due to the qualitative nature of most "performance" criteria for real-time systems, we shall introduce the term "Quality of Service" (QoS) for them.

To continue, some guidelines for efficient QoS-oriented design and development of real-time systems will be given mapping the QoS criteria considerations of the preceding chapter onto the individual phases of a project's life-cycle.

In the chapter to follow, we shall come up with an acceptance scheme for real-time systems comprising the relevant QoS criteria. They address the concepts described in the subsequent chapters and point out their importance to individual QoS parameters.

Since the means to achieve their fulfilment mostly originate from the design domain, in the following two chapters we shall consider "design for QoS" in concordance with the appropriate standards and guidelines. Since the Unified Modelling Language (UML) is becoming a de facto standard also for designing embedded control systems, we shall take into account the pertaining features and profiles together with appropriate QoS guidelines to be addressed in the UML project life-cycles.

Finally, we shall consider real-time standards and system certification. Some technical standards, mostly used in the automotive and telecommunication industries, will be mentioned. Certification will be observed from the point of view of the ISO 900X standards. In particular, we shall deal with the integration of the Capability Maturity Model for (real-time) software into the design and development process of real-time systems. From the certification point of view, also some existing standards for evaluation and performance measurement will be explored.

This book is not intended to be an encyclopaedia containing all knowledge on the subject area covered. It should guide the readers to approaches suitable to design real-time systems with QoS in mind, to cope with implementation or analysis problems and to references on where to look further and/or more deeply. Different groups of readers may use the book differently. Here are some typical use cases:

Students may read the book from the beginning to the end or look up the table of contents for the topics of their interest.

Designers are likely to use the book as a reference guide and to look up individual chapters/sections based on the nature of the systems/problems they are involved in building/solving. The design and evaluation guidelines may be a good starting point for them.

Analysts may first classify the systems they are working on and, then, use the book as a reference guide by working backwards from the flowcharts to the methods and standards listed in the indexes. A good starting point for them are the QoS evaluation decision diagrams.

Researchers can obtain encyclopaedic knowledge of real-time systems and their QoS from this book, offering them a classification of real-time systems and their QoS properties first, and then guiding them through the pertaining methods and standards.

Finally, it is important to note that, unless stated otherwise, all criteria and measures are meant for the system level, i.e. they pertain to hardware and software. The book builds on the presumption of a clear goal and its conscious and systematic fulfilment in correspondence with best practices also mentioned in the book.

2

Performance Metrics for Real-time Systems

So-called benchmark programs or program sets [62, 70, 75] have been selected or designed for the purpose of testing and/or measuring the performance of computing systems in one way or another. They should assist in design or purchasing decisions and in the testing or acceptance stages. Benchmarks are, however, infamous for being both controversial and subjective.

Different kinds of real-time benchmarks have been developed for different areas and types of use:

High-performance computing systems (e.g. signal, voice and image processing, artificial intelligence or numerical analysis)
High-availability computing systems (e.g. control systems, avionics, database or telecommunication systems)

The magnitude of time granularity varies depending on the field of use and type of application. Also, structural awareness of the benchmark application is more important for high-availability computing systems and is present in their benchmarks.

In the sequel, first some background information on performance measurement is given, followed by a selection of some standard benchmarks, which have been devised to evaluate real-time systems. They shall be classified as benchmark programs, timing analysers and performance monitors.

2.1 Benchmarks and RTOS Standardisation

Devising benchmark programs for real-time systems is difficult due to the late standardisation of real-time operating systems. Many different real-time operating systems are currently being used, each one supporting its own specific system calls. The portability of real-time benchmark programs is, therefore, very limited. Some years ago, the set of POSIX 1003.4 standards [7] was defined for real-time operating systems. This should assist in building real-time operating systems with standard interfaces and standard benchmark applications.

To give an example, in [53] a set of benchmark criteria relevant for POSIX systems was defined. These criteria are subdivided into two categories:

1. Those that measure the determinism of the operating system (called "deterministic benchmarks")
2. Those that measure the latency of particular important operations (called "latency benchmarks")

These benchmarks test core operating system capabilities and are independent of any actual application. "Deterministic benchmarks" measure the response time of an operating system, determine if a thread can respond in a deterministic fashion and measure the maximum kernel blocking time. 'Latency benchmarks", on the other hand, are designed to measure the context-switching time between threads and processes, the possible throughput of data between processes and threads and the latency of POSIX real-time signals.

2.2 DIN 19242 Performance Measurement Methods

The German Standardisation Institute's DIN 19242 standard (Part 1) [12] defines performance, measurable parameters and measuring methods for process control computers. Based on these, in [73, 4] a test method and 14 programming examples in C, FORTRAN, Pascal and PEARL are given to measure the parameters defined in DIN 19242. The test suite contains program segments for the following functions: multiplication, conditional branch, sine calculation, matrix inversion, bit pattern detection, interrupt handling with program start, I/O operation on peripheral storage, digital I/O, dialogue response at a terminal, message-oriented synchronisation between two tasks, record-oriented reading from and writing to a (random access) file, data transmission with an error-free transmission circuit and generation of executable code with a subsequent program start. The suite was employed in [8] to compare the performance of a personal computer with that of a microVAX II based on representative performance criteria and appropriate loads. Later, the suite was included in the successor parts of the mentioned standard where, based on applications, appropriate test programs were grouped. In DIN 19243 [13] standard finally measurement and control methods, basic functions of automation with process control computers, appropriate analogue and binary quantities, as well as data acquisition, processing and output were defined.

For a broader application area the DIN 66273 [16] standard was written, which defines data processing tasks, measurement methods, and rating of data processing performance of computing systems, whereby it concentrates on sensible and measurable performance quantities (e.g. response times).

The description of the method to measure execution time according to the DIN 19242 standard is given here as an example. It is based on the following presumptions (see Fig. 2.1):

- Any process control computer runs a multitasking operating system
- There are two programs competing for the CPU — a low-priority process executing a CPU-intensive task that does not require additional resources, and a high-priority process executing the measured task hereby blocking the execution of the former
- The low-priority task's execution time (t_{LP}) is delayed for a time interval being as long as the execution time of the (high-priority) measured process (t_{HP})

The measurement method is the following. First, only the execution time of the low-priority process is measured, then the execution time of both processes is measured (T). The difference of the measured execution times ($T_{meas.}$) represents the execution time of the high-priority process, which also includes the necessary context switches.

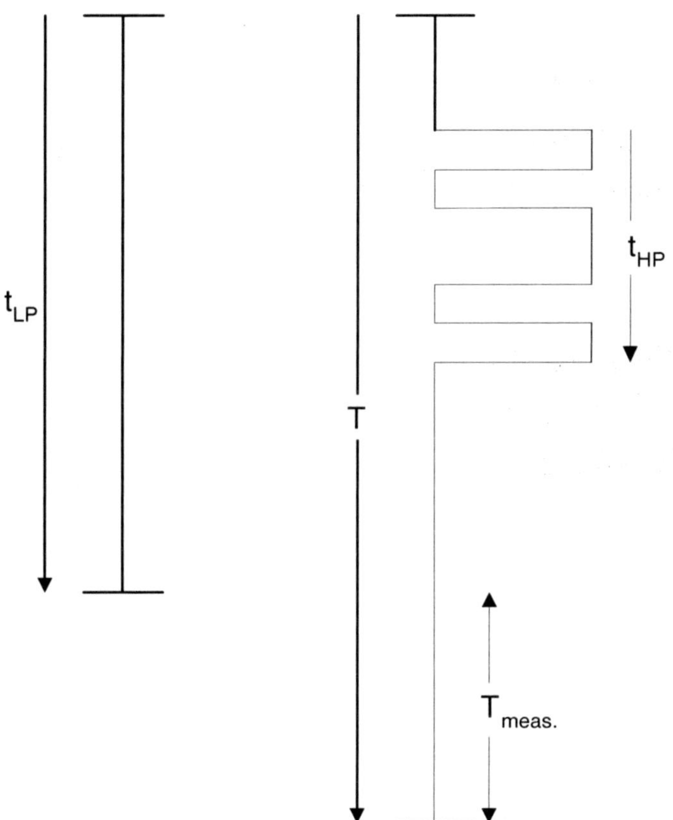

Fig. 2.1. Execution time measurement based on DIN 19242

2.3 Benchmark Programs

2.3.1 Rhealstone Benchmark

The Rhealstone benchmark [42, 43, 44] is a set of six C programs, with which the following characteristics of real-time operating systems can be measured:

Average task switch time: average time for switching between two independent active processes (tasks) of the same priority (i.e. effectiveness of (re-) storing process contexts)

Average pre-emption time: average time for switching to a previously inactive higher-priority process (i.e. effectiveness of schedule-dispatch algorithm)

Average interrupt latency: average time between the occurrence of an interrupt at the CPU and the execution of the corresponding interrupt service routine's first instruction (i.e. effectiveness of dynamic priority assignment and scheduling)

Semaphore shuffle time: average time between a request to lock a semaphore, which is locked by another process, and the time when its request is granted

Deadlock break time: average time to resolve a deadlock that occurs when a lower-priority process holding a resource is pre-empted by a higher-priority process also requiring the same resource

Inter-task message latency: average delay that occurs when sending a message between two processes

The measured times are joined in a common mean value $t' = \frac{1}{6}(t_1 + t_2 + t_3 + t_4 + t_5 + t_6)$ expressed in seconds. The so-called Rhealstone value is defined as the inverse of this value: $\frac{1}{t'}$. Hence, the comparative performance of a real-time system is greater if its Rhealstone value is greater.

The major drawbacks of the Rhealstone benchmark are: first, it measures average times instead of the more important worst-case values and, secondly, the Rhealstone benchmark states that these acquired times should be combined into a single figure by the weighted summation of the measured values. No details are given as to determine the weight factors. More criticism was pronounced by Kasten [44].

2.3.2 MiBench

MiBench [23] is a set of 35 applications typical for embedded systems and divided into six suites, each one targeting a specific area of the market for embedded systems: (1) automotive and industrial control, (2) consumer devices, (3) office automation, (4) networking, (5) security and (6) telecommunications. All programs are available in standard C source code, and portable to any platform that has compiler support. Where appropriate, MiBench provides small and large data sets. The small data set represents a lightweight application of the benchmark, whereas the large data set provides a real-world application. While sharing some common properties with the EEMBC [18] suite, MiBench comes with freely available source code. For hard real-time systems, only the first suite is useful.

2.4 Timing Analysers

To determine timeliness, timing analyses are carried out either off-line or on-line. While off-line timing analysis mainly concentrates on profiling compiled code to discover bottlenecks and to determine a priori timeliness, for on-line timing analyses execution times of the actual application programs or of sets of benchmark programs with known properties are measured.

Owing to the complexity of contemporary real-time applications and processors used, it is difficult to determine exact program execution times off-line — even on the assembly language level. It is necessary to limit the use of high-level language constructs considerably, and to preferably use simpler microprocessors without caches or pipelines in order to be able to obtain useful results. Otherwise, results will be either too pessimistic, or applicable in special cases or for smaller subsystems, only. Hence, off-line analysis has been found to be of limited use for contemporary real-time applications and, although not as precise, overall on-line timing analysis is predominantly employed.

2.4.1 Hartstone Benchmark

The Hartstone benchmark [1] is a set of operational requirements for a synthetic application used to test hard real-time systems. It was written in Ada at Carnegie Mellon University.

The Hartstone benchmark requirements define several test series of processes, viz., periodic with hard deadlines, aperiodic with soft or hard deadlines and synchronisation processes without deadlines. Critical for the Hartstone benchmark are the execution times of synchronisation processes, servicing periodic and aperiodic client processes, which are the sums of the execution times of the critical sections and the independent execution times of the server processes. Each test in a series either meets its deadline, or misses one or more deadlines, and thus fails. There are five test series:

1. Periodic tasks, harmonic frequencies
2. Periodic tasks, non-harmonic frequencies
3. Periodic tasks, harmonic frequencies with aperiodic processing
4. Periodic tasks, harmonic frequencies with synchronisation
5. Periodic tasks, harmonic frequencies with aperiodic processing and synchronisation

For each test series there exists a "baseline task set" which exactly defines the computational load, timing requirements, and period lengths. A test begins with its baseline set, and must fulfil the specified timing requirements. In any subsequent test a single parameter of the tasks in the set is changed. A test is repeated until a "stop condition", usually associated with a deadline or response time miss, is reached. The results of a test series are the task set parameter settings at the time when a stop condition is reached with the highest computational load. As a synthetic computational load the Whetstone benchmark [11] may be used.

Carefully interpreting the results obtained with this test series, one should be able to consider the suitability of a complete platform (hardware and system software) for real-time applications [77]. While the described Hartstone benchmark was devised for single-processor systems, it was later enhanced for distributed environments [41] taking also processor assignment strategies and interprocessor communication into consideration.

2.4.2 SNU Real-time Benchmarks

One of the more recent benchmarks was developed at Seoul National University [68]. It has been devised to run without an operating system. All its functions are provided in the form of C source code. The benchmark programs fulfil the following structural constraints: no unconditional jumps, no exit from loop bodies, no switch statements, no while constructs, no multiple expressions joined by conjunction or disjunction operators and no library calls. As a compilation of some numerical calculation programs (binary search, Fibonacci series, insertion sort, etc.) and digital signal processing algorithms (CRC, FFT, JPEG, etc.) the benchmark is to be used for worst-case timing analysis. Its benefit is independence of target platforms, and its drawback is the limited usefulness in more complex environments with operating systems. From the foreseen experimental set-up (see Fig. 2.2) it was also possible to classify the benchmark programs as the software monitors to be discussed in the next section. The "modified LCC" and "timing analyser" are the core components.

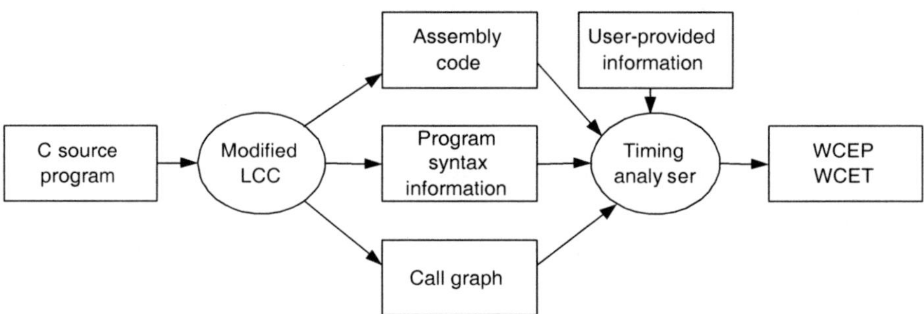

Fig. 2.2. Experimental environment of SNU benchmark for worst-case execution time analysis

2.5 Performance Monitors

Some reasons for monitoring data on the performance of a real-time system are to determine workload characteristics, to verify correctness and to determine reliability. Performance monitoring consists of gathering data on a system during its operation and condensing them into a form suitable for interpretation. Monitoring tools fall into one of the four categories: hardware monitors, software monitors, firmware monitors and hybrid monitors [48]. One may consider most of the mentioned benchmarks as software monitors. In order to obtain more precise results, one may introduce external monitoring devices (hardware monitors), or observe timing events at a lower level (firmware monitors). When hardware monitors have some internal logic for the interpretation of measured times we speak about hybrid monitors.

2.5.1 PapaBench

The PapaBench suite [52] is a real-time benchmark freely available under the GNU General Public License. It is based on the *Paparazzi* project that represents a real-time application developed for different Unmanned Aerial Vehicles (UAV). The benchmark was designed to estimate Worst-Case Execution Times (WCET) and to analyse schedulability. Its programs are simple in structure. With the exception of two while loops, they are composed of not nested for loops with fixed upper bounds, only. Thus, WCET analysis is simplified, and WCET estimation can come closer to the real WCET.

2.5.2 RT_STAP Benchmark

The "Real-Time Space-Time Adaptive Processing" (RT_STAP) benchmark [10] is used to evaluate the application of scalable high-performance computers to real-time implementations of space-time adaptive processing (STAP) on embedded platforms. The STAP techniques are used to support clutter and interference cancelation in airborne radars. RT_STAP is an example of a compact application benchmark that employs a real-time design-to-specification methodology. The scalability study outlined in this benchmark varies the sophistication and computational complexity of the adaptive algorithms to be implemented. RT_STAP provides hard, medium and easy benchmark cases based on three post-Doppler adaptive processing algorithms: higher-order Doppler-factored STAP, first-order Doppler-factored STAP and post-Doppler adaptive Displaced Phase Centre Antenna.

2.5.3 Benchmarks Oriented at Databases and Telecommunication

As compared to traditional benchmarks for On-Line Analytical Processing (OLAP) applications, originating from banking and business applications (e.g. TPC-B, TPC-C), telecommunication applications (e.g. GSM, ATM communication networks) deal with smaller amounts of data. Since these data mostly fit nicely into the main memories of computers, the mentioned traditional benchmarks are not suitable, as they focus mainly on the performance and efficiency of disk access (e.g. TPC-C) and do not consider service times as correctness criteria.

Hence, for telecommunication applications new benchmarks (e.g. [49]) have been developed based on the mentioned ones (e.g. TPC-B) and on predispositions. Here, a similar approach as for database applications is chosen. Test transactions are defined, which are performed in a database of some service provider(s). The Atomicity, Consistency, Isolation and Durability (ACID) properties of transaction processing must be assured by a system under test while the benchmark is being run. The metric's value is reported in transactions per second (*tpsT*), i.e. the total number of successfully committed transactions divided by the elapsed time of the measurement interval. A committed transaction is only considered successful if it has both started and completed within the measurement interval and before its deadline. The transaction success ratio (*successT*) is the total number of successfully committed transactions divided by the total number of transactions entered to a system under test during the measurement interval. The transaction miss ratio (*missT*) then equals $1 - successT$. It represents all unsuccessfully ended transactions.

2.6 Concluding Remarks on Test Methods and Benchmarks

The general division of benchmarks for the two main streams of real-time applications holds for most cases. We may say that application-oriented high-performance benchmarks generally offer more realistic results, since real applications do not differ much from the way the benchmarks use the basic operations. Structurally oriented high-availability benchmarks, on the other hand, also speculate on different scenarios within the applications in which they use the basic operations of the systems under test. Since two applications are never the same, the results obtained adhere more to the properties of the systems than the applications. However, one may also speculate on the relevant results based on the knowledge of a benchmark's structure and the known structure of an application, or even integrate the application into the benchmark. This would, of course, produce entirely accurate results, but is not always possible.

Assessing the capabilities of the above mentioned benchmarks and monitoring methods yields the conclusion that all of them are unable to render the results really needed, because the speed with which a certain arithmetic operation is performed is as insignificant for the evaluation of a real-time application as a whole as, say, an average task switching time or data obtained from test runs, because the latter usually cannot cover all worst-case conditions. Therefore, in the next chapter we take a more thorough look at the fundamental principles of real-time computing to determine performance criteria that really matter.

Although benchmarks have a somewhat bad reputation, it would be good for the real-time community if a group were formed, comprising both system vendors and users, to specify, promote and report benchmarks. System vendors should feel motivated to participate, since this would provide them with an opportunity to spotlight their systems. Users already investing resources in benchmarking systems would feel more motivated by investing in an industry effort that would enable them to influence and leverage the results, ultimately reducing the overall efforts required to accurately assess the desired properties of their computing systems (components). This approach was already successful with the DIN 66273 standard.

3

QoS Criteria for Real-time Systems

There are many questions one may ask oneself when discussing the quality of a real-time system. Generally, there are two typical kinds of questions and answers: final and non-final. For instance, is the execution behaviour of the system (fully) predictable, i.e. is this behaviour known in all situations? If the answer is yes, then this obviously contributes to the system's quality. If the answer is no, we may disqualify the system. Then, for example, we may inquire if the real-time system considered is fault-tolerant (see Fig. 3.1). If it is, we may proceed with our questionnaire. In the case of a negative answer, the question remains as to whether the system endangers its environment upon breaking down. If the answer to this is affirmative, then obviously the quality of the system is insufficient and it should not be used. Otherwise we may proceed with additional questions relevant to this issue or with the next topic (e.g. graceful degradation).

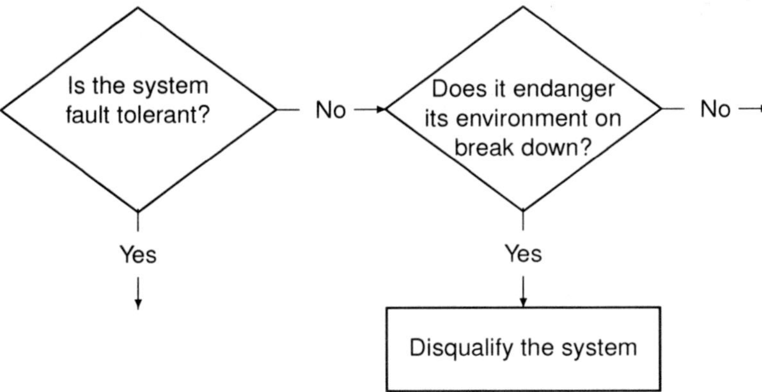

Fig. 3.1. Decision tree to accept or reject a system for a given application

Other questions that follow from the above strategy are: Does the system shut down in a safe and controlled way? Can the system be re-configured or updated during operation? If the answer is no, and the system may never shut down after start-up, e.g. in a space station, then the system is clearly not usable for continuous operation. Do situations exist in which deadlines are missed? If so, are the deadlines hard or soft? A hard real-time system must never miss deadlines. So, if an affirmative answer is given, the system must be disqualified. Is the worst-case performance of the system known? Is it accounted for? Does the system perform correctly, i.e. does it conform to the given functional specification? Are there highly accurate, high-resolution time references? Can the residual and entire execution times of tasks be calculated, predicted or measured? Do synchronisation operations have time-out mechanisms? Are there feasible scheduling algorithms for a given problem? Does the programming language have explicit constructs to express — with unambiguous semantics — time-related behaviour? Can tasks be started, stopped, suspended and resumed at any time? How do the different processors in a distributed real-time system communicate? What is the worst-case time for a message to travel between processors?

As can be seen, there are many questions, and they are difficult to just keep in mind. Hence, they are dealt with in more detail and systematically in the forthcoming sections where also checklists are given for systematic design, development and evaluation of real-time systems. Based on the considerations about the nature of real-time systems from the previous sections, we are now able to state a number of appropriate Quality of Service (QoS) criteria, which can be divided into three groups:

Qualitative exclusive (X group): criteria either being fulfilled or not
Qualitative gradual (G group): one system may have a property to a higher degree than another one, but the property cannot be quantified precisely
Quantitative (Q group): criteria giving rise to measurable numbers

It is important to note that the criteria cannot be used in a general setting, but *only in an application-specific way*. The classification of quality criteria, as shown below, holds for hard real-time systems and applications, and may change for different types of applications (e.g. in soft real-time applications timeliness becomes a quantitative criterion in the form of task completion tardiness).

3.1 Breakdown of QoS Criteria

The quality of a real-time system needs to be assessed by a holistic examination of the entire system encompassing architecture, operating system, scheduling strategy, programming language and design methodology. The quality lies in the extent to which the system fulfils the user's expectations. The following criteria were identified as appropriate and then properly grouped:

1. Qualitative exclusive criteria:
   - X0. Timeliness — the ability to meet all deadlines
   - X1. Functional correctness
   - X2. Permanent readiness — non-stop timely availability of correct service
   - X3. All applicable physical constraints met
   - X4. Licensability — certification (e.g. safety licensability)
2. Qualitative gradual criteria:
   - G0. Timeliness
   - G1. Availability — the proportion of time a system is in a functioning condition
   - G2. Reliability – the degree of a system's ability to perform its required functions under stated conditions for a specified period of time
   - G3. Safety — the degree of protection against risks of errors
   - G4. Security — the degree of protection against consequences of failures being induced from the environment of a system
   - G5. Integrity — the degree of absence of improper system state alterations; further being determined by:
   - G5.1 Robustness — the degree of a system's resilience under stress and
   - G5.2 Complexity (simplicity) — the complexity of a problem is the amount of resources it takes to solve an instance of the problem as a function of the input's size
   - G6. Maintainability — the degree of ability to undergo repairs and modification; further being determined by:
   - G6.1 Portability — the degree of effort needed to port a (software) system to another environment or platform and
   - G6.2 Flexibility — the degree of a system's adaptability to a new environment
3. Quantitative criteria:
   - Q0. Timeliness
   - Q0.1. Execution time benchmarks
   - Q0.2. Worst-case response times to occurring events
   - Q0.3. Worst-case times to detect and correct errors
   - Q0.4. Mean Time Between Failures (MTBF), Mean Time To Failure (MTTF) and Mean Time To Recover (MTTR)
   - Q1. Signal to Noise (S/N) ratio and noise suppression
   - Q2. Capacity reserves
   - Q3. Overall project costs ("the bottom line")

3.2 Concluding Remarks on QoS Criteria

The quality of a real-time system needs to be assessed by a holistic examination of the entire system encompassing architecture, operating system, scheduling strategy, as well as programming language and its compiler; it lies in the extent to which the system fulfils the user's expectations.

Traditionally, quantitative measures form the most popular aspect of performance analyses. Hence, it may come as a surprise that just a few of the criteria compiled above fall into the category of quantitative ones. Care must be taken when employing the measures MTBF, MTTF and MTTR as, for example, a mean time between failures of 100,000 hours does not exclude a break-down during that time.

Criteria relating to implementation artifacts or technicalities have been omitted in favour of those that matter on the user level. For instance, the measures task switching time, pre-emption time and interrupt latency were excluded, because only response time is an application-oriented performance indicator that is important to users.

The assessment of qualitative criteria and measures resulted in questionnaires and decision trees, which will be described in the final chapter. An example thereof is shown in Fig. 3.1. These flowcharts lead to either acceptance or rejection of systems (solutions) for given applications.

4

QoS-oriented Design and Evaluation of Real-time Systems

In this chapter, we consider the life-cycle of projects realising real-time systems. Here a "roadmap" is given for the QoS criteria defined, with references for further reading in the pertaining chapters and sections of this book. For easier comprehension it is to be noted that *predictability* and *dependability* are to be understood as collective notions of *timeliness* and *functional correctness*, and of *availability*, *reliability*, *safety*, *security*, *integrity* and *maintainability*, respectively.

Table 4.1. Overview of QoS criteria and measures

Exclusive QoS criteria	
X0 (timeliness, Sect. 5.2.2)	S,D,I
X1 (functional correctness, Sect. 5.2.1)	S,D,I
X2 (permanent readiness, Sect. 5.2.3)	S,D,I
X3 (physical constraints met, Sect. 5.2.4)	S,D,I
X4 (licensability, Sect. 5.2.5)	S,D,I,V
Gradual QoS criteria	
G0 (timeliness, Sect. 5.3.1)	S,D,I,V
G1 (availability, Sect. 5.3.2)	S,D,I,V
G2 (reliability, Sect. 5.3.3)	S,D,I,T,V
G3 (safety, Sect. 5.3.4)	S,D,I,T,V
G4 (security, Sect. 5.3.5)	S,D,I,T,V
G5 (integrity, Sect. 5.3.6)	S,D,I,T
G6 (maintainability, Sect. 5.3.9)	S,D,I,T,M
Quantitative QoS criteria	
Q0 (timeliness, Sect. 5.4.1)	S,D,T,V
Q1 (S/N ratio, Sect. 5.4.2)	S,D,T,V
Q2 (capacity reserves, Sect. 5.4.3)	S,T,V
Q3 (overall costs, Sect. 5.4.4)	S,V

In Table 4.1 the QoS criteria listed are set into relation with the measures defined for different phases of the real-time system life-cycle. The phases of the life-cycle are abbreviated as follows: S (specification), D (design), I (integration and implementation), T (verification), V (validation and licensing) and M (updating and maintenance). In the sequel, the individual phases are discussed in more detail.

4.1 Introducing QoS Parameters into the Design of Real-time Systems

During a project's life-cycle, the qualitative exclusive (X group) criteria are fulfilled in the phases of specification and construction of a conceptual design model. Here, partly also implementation issues are considered, mainly by the choice of appropriate methods and mechanisms to provide for, e.g. permanent readiness or licensability. The gradual (G group) criteria are satisfied on different levels of design and implementation, in the course of specifying the properties of constructs and refining the system model. They are evaluated after individual phases of design and implementation during verification of the system model designed. The quantitative (Q group) criteria, although pertaining to the previously mentioned phases from refinement of the system model through implementation, are evaluated last, viz., during model analysis and its validation, after the model is deployed and realised on a target platform. The methodical support to satisfy the QoS criteria introduced is outlined in the sequel.

The life-cycle of real-time (embedded) systems is usually rather specific due to the great diversity of their applications, although it also contains the mentioned phases common for information systems. As hardware and software need to be matched perfectly, they are usually designed conjunctively, i.e. co-designed. Since best results, with respect to QoS, are achieved by employing this approach, the description of applying the mentioned criteria and of methods for satisfying them will follow these phases.

4.1.1 Hardware and Software Architecture Modelling

Hardware and software architectures are defined by determining the properties and structure of the corresponding components as well as by defining the mappings of software to hardware components, and possibly also re-configuration scenarios.

Timeliness (X0), being the capability to meet all deadlines, is supported in design methodologies by the ability to reason about timing limitations at early stages in order to discover specification errors or bottlenecks as soon as possible. Later, timeliness/schedulability analysis of the models designed usually provides more accurate timing information. This stage, at which major re-design decisions are much harder, should, however, mainly be dedicated to design fine-tuning.

Simultaneous operation towards the outside (timeliness – X0) and physical constraints (X3) define environmental conditions and limitations together with real-time constraints on "performance" or, more precisely, the physical interface and timing constraints of the environment on the individual operations of an embedded system. By observing stimuli from the environment and the response of embedded systems to them, it is usually determined whether deadlines are met and how long the response times required are. Timing parameters are taken into consideration during model analysis by co-simulation and possible later schedulability analysis, which provide a good foundation for the implementation phase. The more accurate a model is the more accurate are the timing results. This, however, does not render timeliness a gradual criterion. It must be sustained during the modelling, analysis and implementation phases in order to provide the QoS requested. Hence, the methodical mechanisms and tools used should support this. Usually, the modelling and evaluation cycles can be repeated, and a model can gradually be improved until the designer is satisfied with the result, viz., when the specified timeliness, functional and physical requirements are met.

Permanent readiness (X2) in addition to timeliness is an option that pertains more to the kind of project than to a model. From the design perspective, it must mainly enable the definition of operation scenarios such as maintenance, error handling, functionality upgrades, or safety and security features and how timeliness is sustained while performing them.

4.1.2 Model Refinement and Checking

Functional correctness (X1) relates to designed system's models and their implementations. Possibly, simulation/execution models are automatically generated from specifications. During design, system models are subjected to the following consistency and coherency checks:

Completeness check ("all components are present and fully described"): The functional correctness of a system depends on the fact whether all applicable physical constraints are met (X3). This applies to the degree to which they can be specified and considered in a model. A system model needs to be complete, i.e. all components must be present and fully described prior to its transformation to the target platform or simulation models. All component data are mandatory, since any lack thereof could represent an inconsistency. Hence, all components have to be checked for the presence of their properties.

Range and compatibility check ("parameter compatibility between related hardware and software components"): Each parameter/property of a component should be range-checked in order to prevent input of erroneous/meaningless data. On the other hand, the compatibility of interconnected components also needs to be checked (e.g. interface compatibility on lines between devices, or data directions between connected ports). Preferably only static structures and real values should be used to provide additional confidence in the models. By itself, however, this does not assure anything, but contributes in combination with other checks mentioned to reliability (G2), robustness (G5.1) and portability (G6.1).

Software-to-hardware mapping check ("complete coverage and consideration of resource limits"): By checking the coherence of software-to-hardware mapping, the absence of software or hardware components can be determined. In doing this, also the interconnections between software components and their reflection by hardware interconnections are checked.

While the absence of a component or its properties is usually determined manually (by inspection or audit), range, compatibility and mapping may also be checked automatically. The goal is to come up with a consistent model, which is correct abreast to the specifications. The subsequent checks for timeliness and performance (Q0) may then reveal possible specification errors.

The automatic creation of models from specifications and their maintenance also helps in managing complexity (G5.2), because models must be complete and consistent while, on the other hand, they should be kept as simple as possible. This again contributes to reliability (G2) and robustness (G5.1). In combination with appropriate methods and measures, however, it also facilitates portability (G6.1) and flexibility (G6.2).

4.1.3 System Model Co-simulation

By co-simulation mainly timeliness (X0, G0), functional correctness (X1) and deterministic behaviour with respect to the specifications are checked. This also contributes to dependability and predictability, with emphasis on reliability (G2), robustness (G5.1) and safety (G3), and provides means for checking a system's safety (G3) with harmless tests of fault tolerance mechanisms, e.g. graceful degradation upon malfunctions. The latter is a safety (G3) feature of a (final/working) system, which tells if it raises an alarm and fails immediately, or tries to "survive" the situation and continues working when the operation mode returns to normal.

While dependability mainly depends on the consistency of the system model and the rigour of the corresponding implementation mechanisms applied, predictability represents the degrees of reliability (G2) and timeliness (G0/Q0). Unwanted unbounded delays or arbitrarily long executions (conflicting with X0 and G0) can be handled on the operating system level, e.g. by preventing deadlocks and infinite execution of single task loops, and by using systematic and rigorous program development methods to produce well formed programs. Potential bottlenecks can be discovered by co-simulation and, then, corrective measures can be taken. Popular RTOS-based deadlock prevention techniques are, e.g. the Priority Ceiling and Stack Resource Protocols to handle deadlocks caused by priority inversion with priority or Earliest-Deadline-First scheduling, respectively. Other sources of non-determinism may also be recognised during co-simulation and removed as errors in design.

Many design methodologies offer a wide variety of constructs (e.g. watchdogs, time-out scenarios, re-configuration options), as well as consistency and coherence checks (e.g. parameter compatibility of related constructs, range checking), which affect timeliness (G0, Q0), safety (G3) and reliability (G2). Security (G4) is also supplemented by these features, although other protective mechanisms have been developed and should be used in aggressive environments (e.g. data encryption, preservation of memory space integrity).

4.1.4 System Integration and Implementation

The implementation phase may logically succeed the co-simulation phase, possibly by using the co-simulation model to the largest possible degree in porting the designed software onto the specified hardware platform. The changes introduced at this stage should be kept to a minimum, and the process should be automated in order to reduce the possibility of invalidating a functionally correct (X1) and temporally consistent (X0, G0, Q0) model.

4.2 System Evaluation

Worst-case response times to occurring events (Q0.2, Q0.3) are usually measured at final systems. Since they are determined in the specification phase, and these criteria are followed throughout the entire life-cycle, traceability is ensured. Only minor deviations from the timing constraints specified are expected. If major deviations occur, however, they can be traced back to their sources (usually inappropriate device parameters, to be corrected and re-considered in co-simulation, which should point to possible solutions) and corrected accordingly. Capacity reserves (Q2) are usually determined throughout timing (Q0) analysis.

Worst-case times to detect and correct errors (Q0.3) are a special kind of (Q0.2), when fault tolerance options of a project are tested. In the initial phase, they can be checked by co-simulation, whereas in the final run they are usually tested by a monitor/test-device being attached to the inputs/outputs of a system.

Signal-to-Noise (S/N) ratio and noise suppression (Q1) is a special combination of (Q0.2) and (Q0.3), when the process of correcting errors is integrated into the main functionality of a system. It addresses the mentioned fault tolerance and graceful degradation features.

Computing MTBF, MTTF and MTTR (Q0.4) is a tedious and lengthy process. Usually, the previously mentioned tests from (Q0.1) through (Q0.3) are executed repeatedly, and their results evaluated statistically to produce the mentioned QoS criterion.

4.3 System Certification

Although some of the mentioned methods are standard, the guidelines on how to use them and check designs are not. So far the closest to quality-oriented standards for (software) systems engineering are Capability Maturity Model Integration (CMMI) and its successor ISO/IEC 15504 [33] on Software Process Improvement and Capability Determination (also known as SPICE) with its associated quality standard ISO/IEC 90003 [38] oriented at software engineering in combination with guidelines for the application of ISO/IEC 9001 [37] to computer software. Although having no special notion of real-time systems, these standards may, nevertheless, be considered to investigate whether projects generally comply with ISO/IEC 9001. Characteristic for real-time projects are the (design) methods used, but not the way they are used. The mentioned ISO/IEC 90003 provides references to related ISO standards (viz., ISO/IEC 12207 [30] for software life-cycle processes with amendments, containing instructions for testing, documenting, etc., from which also ISO/IEC 15504 has been derived) and guidelines on how to apply them in combination.

By following the standard design and development guidelines, one could license a final system, as being developed in concordance with the pertaining standard. If the system's environment also operates under the restrictions of the same standard, the matching quality systems render a perfect combination, also resulting in a certificate (license) on standard compliance of the (business) process.

Hence, we may say that licensability (X4) pertains more to the systems designed in concordance with the regulations specified than to the choice of methods used to develop them. It is up to a method, however, to provide means and certified testing environments to test products. Guidelines according to IEC 61508 may be considered when discussing e.g. safety-licensing.

In any project the overall project costs (Q3) are the "bottom line". They pertain to combined cost of hardware and software as well as to their development. Except for the fact that complex custom designs may significantly increase the costs of the hardware parts of systems, this book makes no further statements on this subject. For both hardware and software it holds that using standard-compliant components decreases overall development time and cost, and facilitates certification at the system level when compatible standards are employed in the development processes.

4.4 Outlook

With the help of the methodical guidelines presented one may efficiently design QoS-aware (embedded) real-time systems. To evaluate their performance, various benchmarks are available and may be used to check whether they meet the given system specifications. However, since these guidelines are non-standard, the approach of DIN 66273 should be considered with appropriate "performance" parameters defined to achieve a certain level of certification.

As a consequence of using the guidelines presented in this book, the quality of design processes as well as of systems designed should be enhanced and their cost reduced, due to making the right design and development decisions at proper times. This should shorten overall development time and allow minimisation of maintenance costs. Last but not least, it should render designed systems safer and more secure.

5

QoS Decision Flowchart for Real-time Systems

In this chapter, the individual QoS criteria are revisited and dealt with in more detail. They are presented in the form of descriptions with associated flowcharts (see Fig. 5.1) and guidelines on where to look next or disqualify a system. First, the qualitative exclusive criteria are considered, since they offer the easiest way of reasoning on rejecting systems in case they do not comply with specified requirements. If the latter are partly fulfilled, the appropriate qualitative gradual or quantitative criteria are checked in order to determine whether the degree of compliance is sufficient or not.

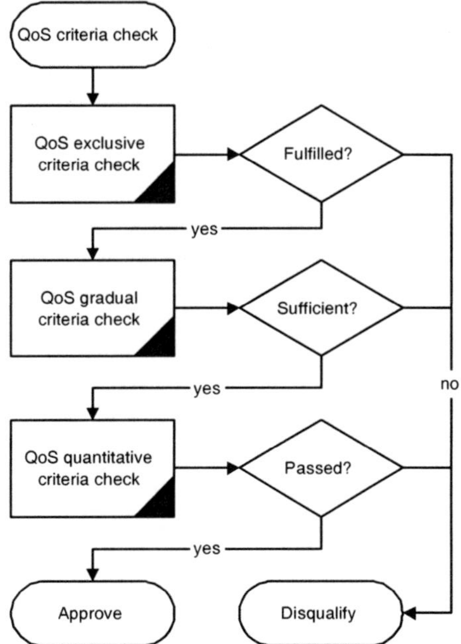

Fig. 5.1. Flowchart to check top-level QoS criteria

The basic two QoS properties of any real-time (embedded) system are its predictability and dependability. A simplified definition, which is often used with real-time systems, defines *dependability* as proper functioning, while *predictability* is usually referred to in connection with their timeliness. All other mentioned QoS criteria support and help to determine the degree of predictability and dependability offered by a system.

5.1 Predictability and Dependability

Predictability: "is a degree of trust in that a correct forecast of a system's state can be made qualitatively or quantitatively." It represents a degree of behavioural determinism and timeliness even in error situations. Functional and temporal predictability are considered here as one integral property.

When discussing a system's predictability it is reasoned upon the portion of cases in which the system behaves as predicted or, when comparing two systems, which one of them behaves predictably in more cases. It mostly concerns timeliness in connection with fault forecasting and prevention measures applied in order to achieve better functional and temporal predictability.

To give an example of an experimental set-up to "measure" and compare predictability, one could monitor (a) concrete application(s) and measure the rates of deadline misses and faults under varying load/scenarios. Then one would compare these values with the specifications.

Similar statements as above apply also to *dependability*, which has already been dealt with in the section on design for dependability. While predictability is mainly addressed by fault forecasting and fault prevention, dependability on the other hand is supported by fault tolerance and fault removal mechanisms.

Dependability: "is a degree of trustworthiness of a computing system, which allows reliance to be justifiably placed on the service — function — it delivers." According to [5], it is considered as the sum of availability, reliability, safety, security, integrity and maintainability. Depending on the application, one might ponder the individual "summands" with respect to their relative importance to the concrete application. These properties, which collectively define predictability and dependability, are dealt with in more detail in the sequel.

5.2 Qualitative Exclusive Criteria

In this section, the qualitative exclusive QoS criteria are dealt with. In Fig. 5.2 the approach to check them is outlined. In the sequel, each of the criteria groups is dealt with in more detail.

5.2.1 Functional Correctness (X1)

Functional correctness is equally important for real-time applications as for any other computational application. It is difficult to prove formally. Often it is checked by

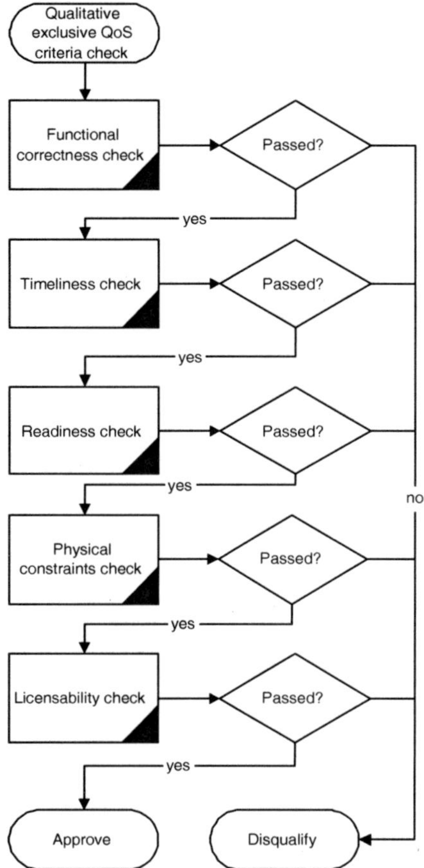

Fig. 5.2. Flowchart to check qualitative exclusive QoS criteria

black-box or grey-box testing of a final product, which helps to detect most functional faults, provided the specification is strict enough. Such tests do not prove the functional correctness of an application, but they are usually sufficient to be able to say that it behaves *correctly in concordance with the specifications*. Making modifications at this point, however, is usually a tedious process and may introduce new errors. Hence, in the scope of the chapter on *Design for QoS* measures for mastering complexity and supporting the suppression of faults were described, which, when used correctly, contribute greatly to functional correctness. Of course, their usage alone cannot prevent errors, but they help profoundly in localising and removing the sources of faults in a systematic way.

Equally important for the correctness of real-time applications are timeliness issues, which shall be dealt with in more detail below. In any case, a functionally incorrect system or application program should be disqualified right away, since by this fact all other issues become irrelevant. For the necessary steps to check functional correctness see Fig. 5.3.

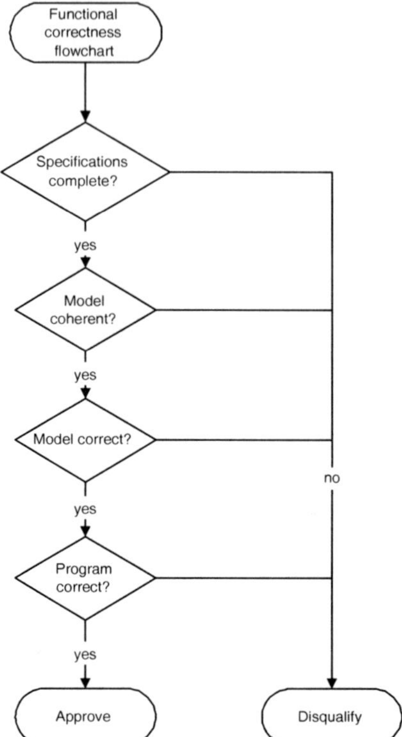

Fig. 5.3. Flowchart to check functional correctness

5.2.2 Timeliness (X0)

Timeliness: "the ability to meet all deadlines." This is an exclusive criterion in hard real-time systems, i.e. the deadlines of a system are either missed or never missed. Systems missing deadlines are not admissible in hard real-time environments.

Like functional correctness, timeliness as such is rather difficult to prove (see Fig. 5.4). Diverse monitors, performance benchmarks as well as timing and schedulability analysers have been developed for this purpose. They act on various levels, from the one of high-level language source code via assembly language to the hardware level, to enable timing analysis. In general, they are not successful in providing exact timing results, since in modern highly parallel processors it is not possible to exactly predict the execution time of instructions, even on the assembly language level. Usually, they offer estimates of the actual execution time that could be considered as upper bounds in most cases, but on the one hand these are sometimes too pessimistic, whereas on the other hand there are exceptions in dynamic execution, especially within RTOS-enabled environments, which could disqualify the results. Therefore, in practice one settles for *timely deterministic behaviour*, which represents the timing equivalent to black-box functional testing. The results are usually provided by monitors in the course of executing test scenarios, measuring execution times and comparing them with maximum reference values.

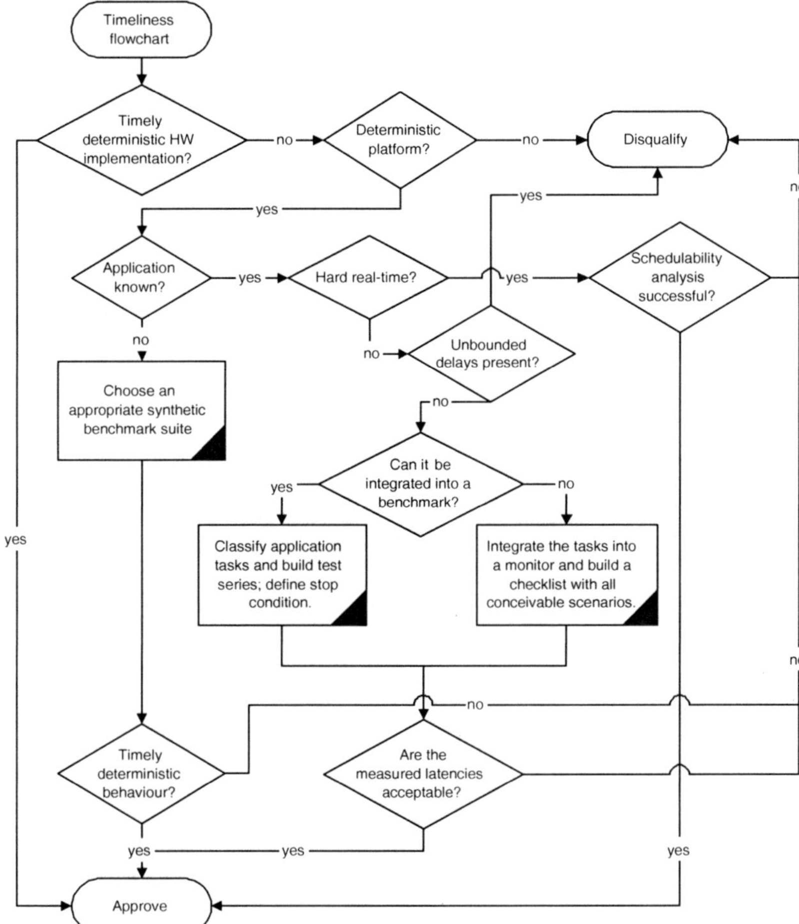

Fig. 5.4. Flowchart to assess timeliness

5.2.3 Permanent Readiness (X2)

Permanent readiness: "non-stop timely availability of correct service." It includes deterministic behaviour, i.e. the absence of improper state changes with no unbounded delays or arbitrarily long executions (deadlocks prevented).

Permanent readiness (see Fig. 5.5) is supported by fault tolerance. Besides high availability, permanent readiness also means that a system's reaction times should be held within prescribed limits during its entire up-time.

One may say that system operation simultaneous with the environment is required to ensure permanent readiness. To facilitate continuous operation, dynamic re-configuration mechanisms should be introduced to enable deterministic addition and removal of system components under well defined consistency constraints, and with minimum and predictable effect on execution time.

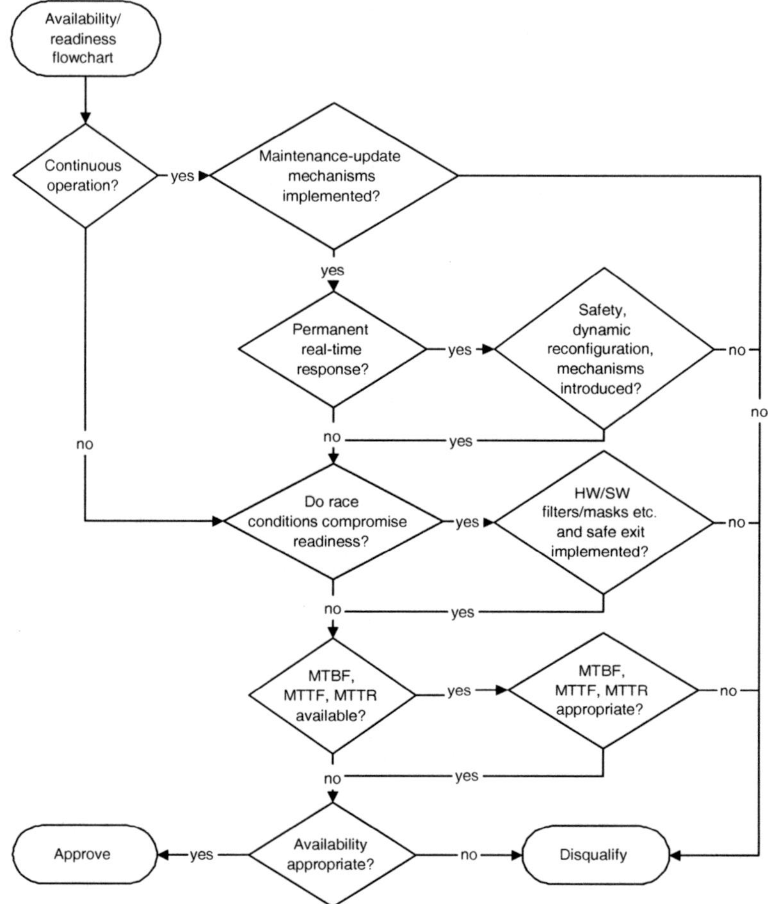

Fig. 5.5. Flowchart to check availability/readiness

5.2.4 Meeting All Applicable Physical Constraints (X3)

Physical constraints (for guidelines see Fig. 5.6) pertain to hardware platforms used, I/O ports and interfaces, as well as to the program structures of software applications. In order to assure that these constraints are met, a perfect match among both system and application components should be established. Possibly, only static data structures and real values should be used. Non-compliance with the mentioned guidelines may introduce safety and security hazards, which have to be addressed accordingly in order to sustain safe and deterministic operation.

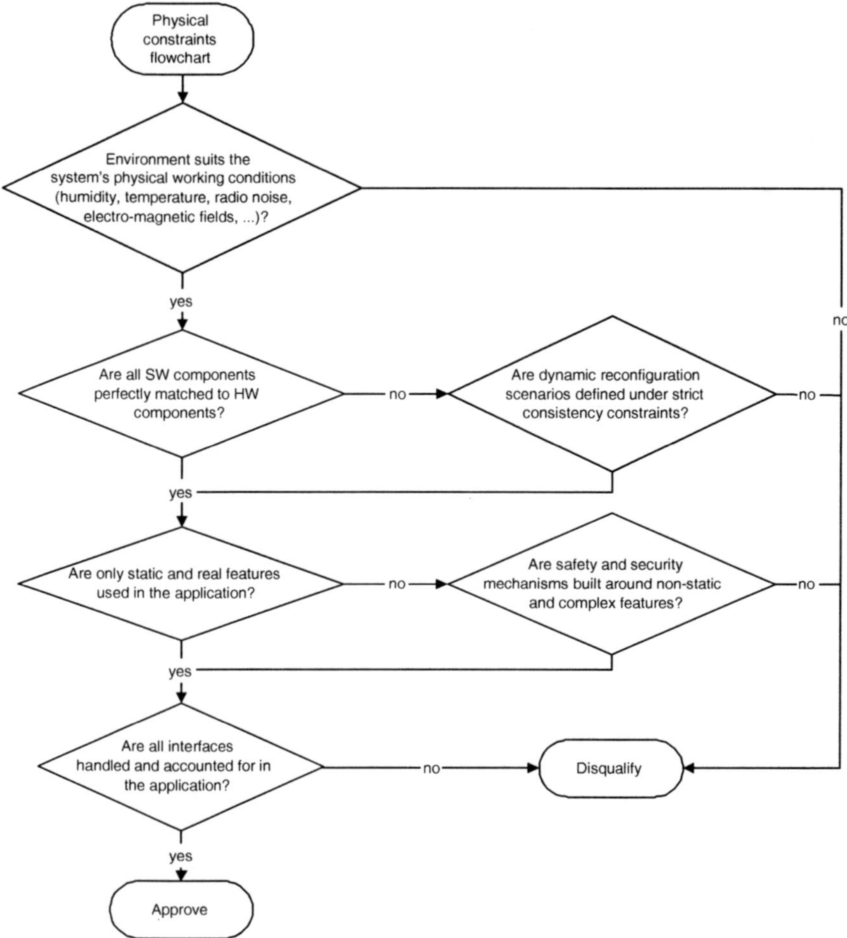

Fig. 5.6. Flowchart to check physical constraints

5.2.5 Licensability (X4)

A system can be certified on compliance to standards or individual QoS criteria, e.g. safety, availability, maintainability, integrity, or security. Standards assume a certain level of provided QoS. Hence, they also prescribe the design methods and development mechanisms that must be used to ensure the required level of QoS.

For a system to be certified for quality it has to be developed in concordance with a standard development scheme like the Capability Maturity Model (CMM) for Software [59], which has to be sustained in the engineering of software projects to fulfil the requirements of the standard ISO 9001. For system level certification one should seek hardware building blocks or bundles thereof from ISO 9001-certified vendors, so that certified components can simply be included in the native quality system. The ISO 9000 standards certify "only" quality-aware system development.

To issue a safety certificate, software products have to be compliant with, e.g. IEC 61508 where a particular level of compliance (SIL) can be determined. They guarantee a system's users/environment a certain level of absence of catastrophic consequences of failures (see Fig. 5.7). .

Similarly, one can discuss security licensability. The applicable security methods and mechanisms used during system design and development are considered to determine the level of security (in correspondence with, e.g. [50] or [9]).

5.3 Qualitative Gradual Criteria

Although we might express some of the gradual QoS criteria analytically as equations and operate with the results statistically, generally they cannot be quantified — mostly they are referred to as *high, medium* or *low*, or we may say that a system (version) fulfils some criteria to a higher degree than another.

In this section the qualitative gradual QoS criteria are dealt with. In Fig. 5.8 the approach to them is outlined. In the sequel, each of the criteria groups is covered in more detail.

5.3.1 Timeliness (G0)

While being an X criterion in the hard real-time domain, for most real-time systems *timeliness* is (also) a G criterion. Hence, depending on the application, it has to be considered on either side or with respective rigour. Based on the X1 criterion, a system is disqualified if a deadline is missed. By applying G0, the system is not disqualified until a certain number of deadlines is missed or a certain time laxity is overrun. One may also compare systems based on their respective deadline miss counts/laxities and rate them based on their comparatively better/worse timeliness.

5.3.2 Availability (G1)

Availability: "is the proportion of time a system is in a functioning condition." It also represents *readiness* (see Fig. 5.5) for correct service or the degree to which a system suffers degradation or interruption in its service to the user as a consequence of failure of one or more of its components. Its exclusive equivalent is *X3 — permanent readiness* (see Fig. 5.5).

A system's availability can be supplemented by *graceful degradation. Graceful degradation upon malfunctions represents a degree of a system's functionality*

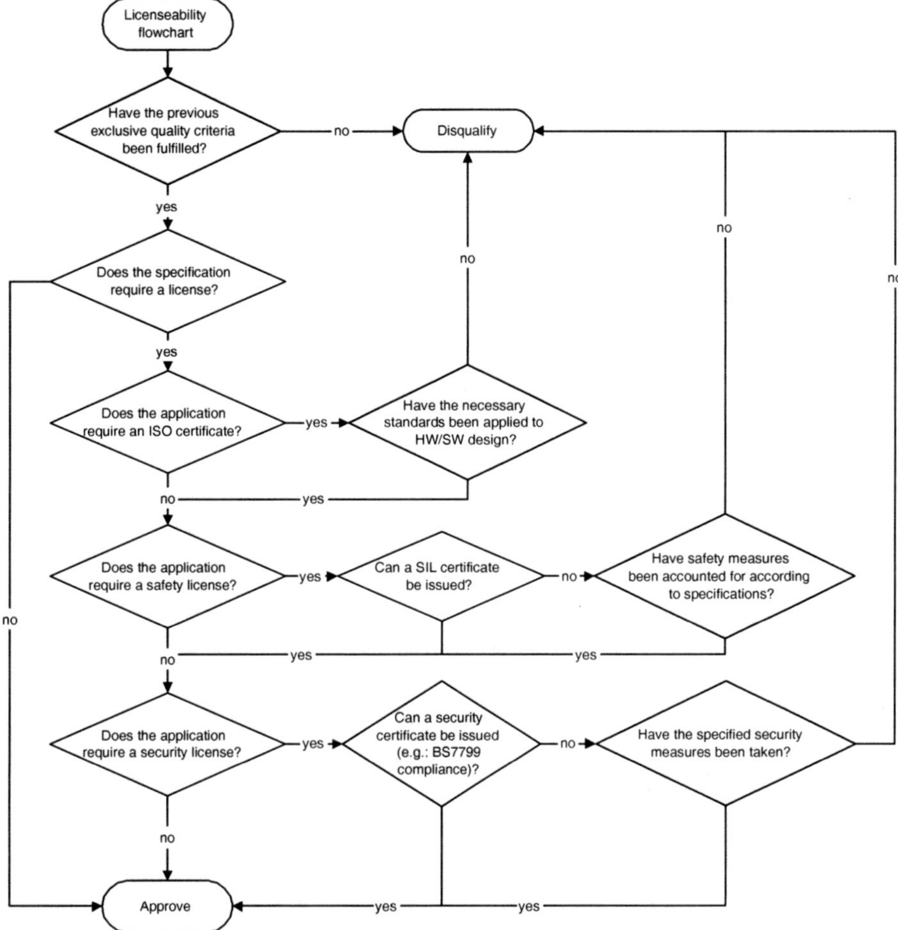

Fig. 5.7. Flowchart to assess licensability

provided in error situations. If the system's operating quality decreases at all, the decrease should be proportional to the severity of the failure. The mechanisms to achieve this originate from the domain of fault tolerance, which was discussed in the section on design for dependability. *Fault tolerance is defined as the degree of a system's provided ability to continue working properly in the event of failure of some of its components.*

5.3.3 Reliability (G2)

Reliability: "is a degree of a system's ability to perform its required functions under stated conditions for a specified period of time." Reliability represents the continuity of correct service (both functionally and temporally) and is dependent on other

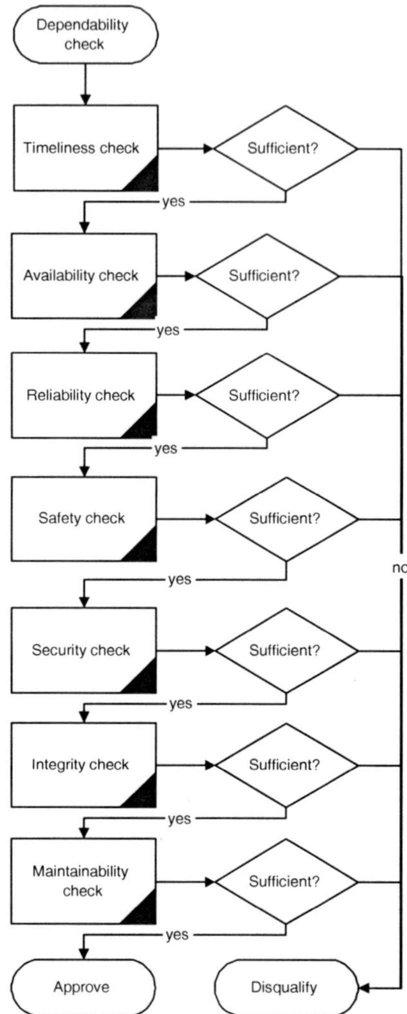

Fig. 5.8. Flowchart to check qualitative gradual QoS criteria

dependability criteria as well. During design it is best supported by *fault prevention*, however during operation, *fault tolerance* mechanisms ensure that it is sustained. Also, during design and operation security methods and mechanisms apply to ensure secure operation.

5.3.4 Safety (G3)

Safety: "represents the degree of protection against risks of errors." It is closely related to *reliability* and *robustness*, which both support a system's safety. A safety-critical application represents an application whose malfunction introduces risk to human lives/health or can cause damage to the system itself and/or its environment.

In the section on safety-oriented design, a number of measures that contribute to system safety was listed. The appropriate questions to be asked here could be formed from a checklist of these measures, namely, which ones have been used or implemented and which ones not (see Fig. 5.9). Depending on the result, a degree of safety can be established (also formally, e.g. in the form of SILs [28]), which may be sufficient or insufficient with regard to the safety requirements from their specifications.

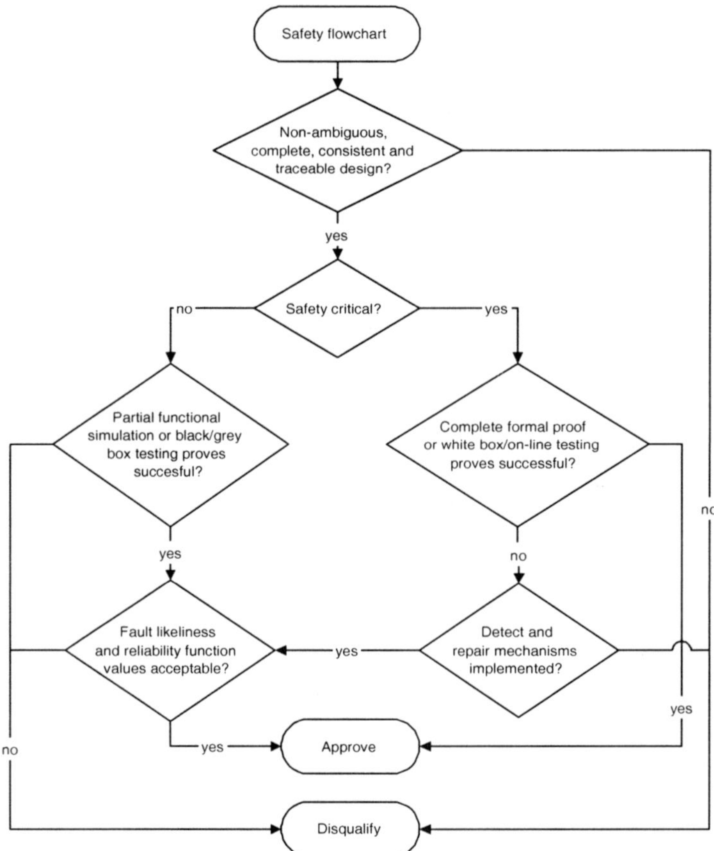

Fig. 5.9. Flowchart to assess safety

To give an example, the already mentioned *safety shell* has been developed to encompass safety-critical real-time applications. Each of its components (state monitor, I/O monitor, timeliness monitor) contributes to an application's overall *safety*, while their absence or reduced functionality diminishes it.

5.3.5 Security (G4)

Security: "is a degree of protection against consequences of failures being induced from the environment of a system." It is seen as a sum [5] of *availability*, *confidentiality*, and *integrity* — in the sense of absence of unauthorised state alterations, where confidentiality represents the degree of protection against unauthorised disclosure of information.

In the section on design for security, a strategy was presented on how to provide a secure environment for an embedded (hard) real-time system. As with safety, the degree of security depends on the methods and mechanisms applied to ensure a system's secure operation. The appropriate questions to be asked here form a checklist of the applicable methods and mechanisms (see Fig. 5.10). Of course, one cannot judge the effectiveness of a system's security solely based on the number or the kind of methods and mechanisms applied. When giving statements on the degree of security, one can only refer to security standards (as listed in, e.g. [50] or [9]).

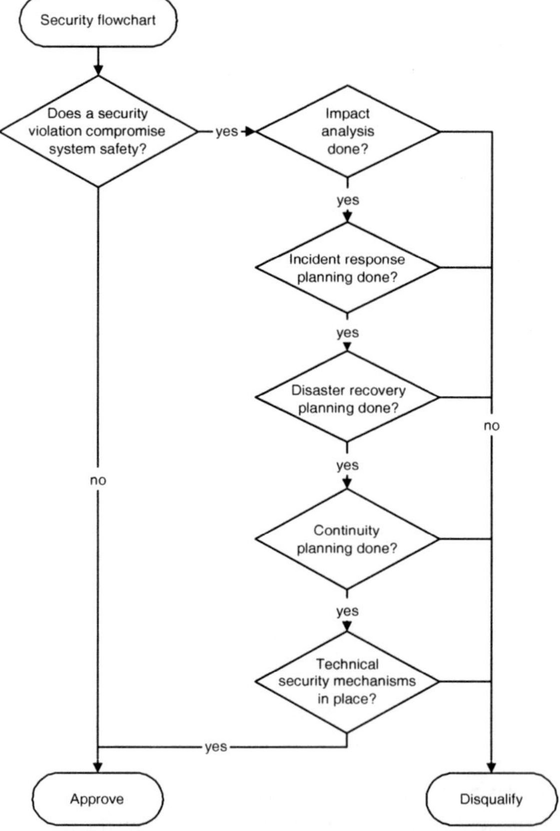

Fig. 5.10. Flowchart to assess security

5.3.6 Integrity (G5)

Integrity: "represents the absence of improper system state alterations." [5] It is prerequisite for safety, availability and reliability, and is rarely considered as a stand-alone attribute. Although the mentioned relation does not hold for confidentiality, one could also consider security in the context of integrity. There is no direct relation among integrity and maintainability, but maintenance should not compromise a system's integrity.

A good example of an environment for ensuring the overall integrity of an application is again the *safety shell*. The *monitors* have the function to ensure the application's integrity. Whenever a system parameter is about to enter *unsafe* or *illegal* value spaces, a corrective action that diverts it back into the *safe area* is initiated.

Integrity is supported by robustness, which is achieved by fault tolerance measures, and simplicity of a system's design (see Fig. 5.11). A complex system design is likely to undermine the system's integrity.

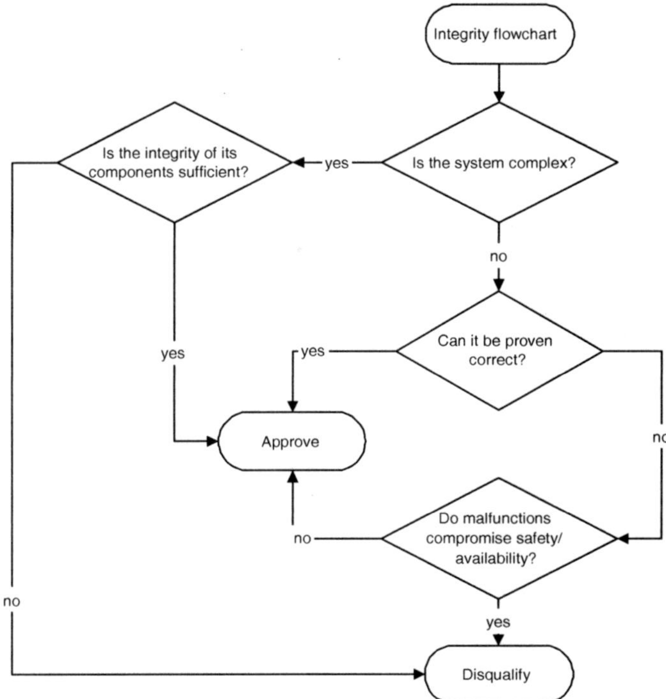

Fig. 5.11. Flowchart to assess integrity

5.3.7 Robustness (G5.1)

Robustness: "is a degree of a system's resilience under stress or when confronted with invalid input or changes in internal structure or external environment." Robustness supports integrity and is closely related to *reliability*. In contrast to reliability it concentrates on the number of changes, fault rate, or number of deadline misses that the system "survives" intactly.

5.3.8 Complexity (G5.2)

Complexity: "in computational complexity theory, the time complexity of a problem is the number of steps that it takes to solve an instance of the problem as a function of the size of the input (usually measured in bits), using the most efficient algorithm — this allows one to classify problems by complexity class (such as P or NP). Such an analysis also exists for space, i.e. the memory used by the algorithm."

There is an entire theoretical field concerned with computational complexity. To concretise the above definition, however, some ideas on how to consider it are given in the sequel.

Means to express complexity are, for example, the number of logical paths through an application program, or the number of states in a program's equivalent implementation as state automaton. One may also see the same criterion in the amount of silicon required for a program's complete hardware implementation (e.g. in an FPGA).

One may say that (unnecessary) complexity generally diminishes integrity while, on the other hand, simplicity (indirectly) fosters it.

5.3.9 Maintainability (G6)

Maintainability: "represents the degree of the ability to undergo repairs and modification." Maintainability is an integral property supported by portability and flexibility (see Fig. 5.12).

5.3.10 Portability (G6.1)

"Portability represents the degree of effort needed to port a (software) system to another environment or platform." Portability is affected by the complexity of a (software) system and the number/complexity of the components that need to be (ex-) changed. It is determined by the number of platforms on which an application can run. This criterion is carried over to *adaptability* in the sense of the number of changes that have to be made in order to be able to port (transfer/translate) the application to another platform (see Fig. 5.12).

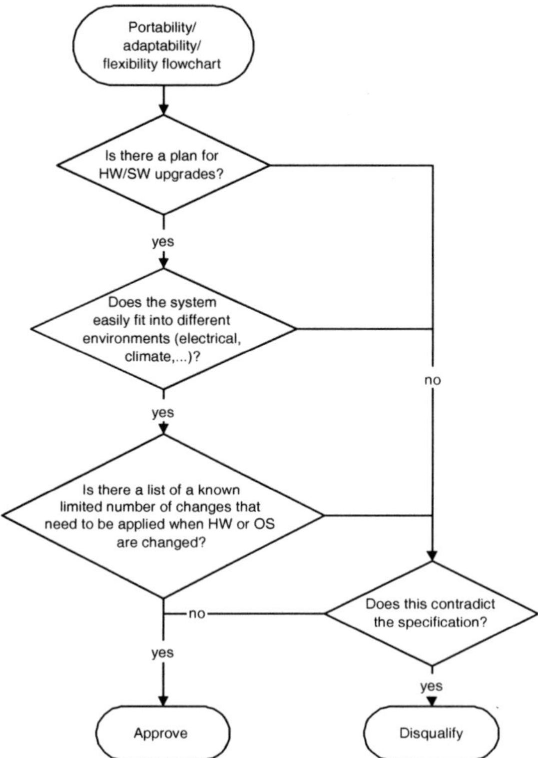

Fig. 5.12. Flowchart to assess maintainability

5.3.11 Flexibility (G6.2)

"Flexibility represents the degree of a system's adaptability to a new environment." We may also consider it as resilience in recovering from a shock or disturbance originating from (a change in/of) the environment.

Similar statements as for portability also apply to flexibility (see Fig. 5.12), which represents a measure for the effort to achieve adaptability regarding portability and subsequent (version) upgrades. The degree of flexibility is highest for applications where portability and adaptability have been accounted for during their design (e.g. on-line/automatic upgrades of functionality or data).

5.4 Quantitative Criteria

In this section the quantitative QoS criteria are dealt with. In Fig. 5.13 the approach to check them is outlined. In the sequel, each of the criteria groups is considered in more detail. In the course of this, also those qualitative exclusive and gradual criteria that share quantitative aspects are addressed again.

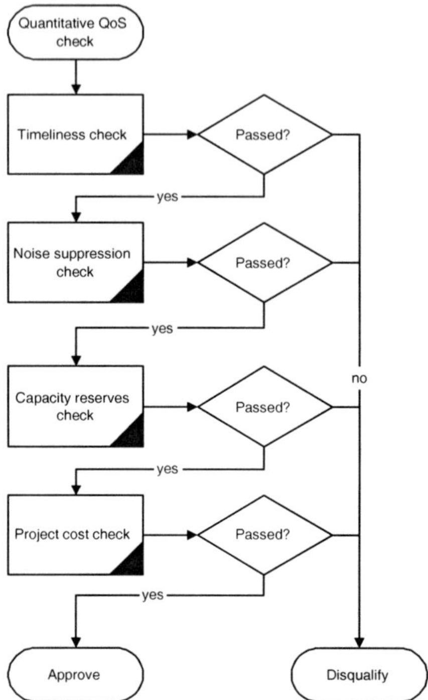

Fig. 5.13. Flowchart to check quantitative QoS criteria

5.4.1 Timeliness (Q0)

Timeliness, being a qualitative exclusive (X0) and a gradual criterion (G0), has quantitative facets, too. In general, timeliness T may be expressed by $T = \frac{E}{B}$, with E representing performance and B load. By performance here functionally correct simultaneous execution is meant, and the amount or frequency of inputs is referred to by load. As soon as the ratio falls below 1, deadlines are missed and it is up to the application's specification to define a lower bound — required "degree of timeliness" — or an upper bound for the "peak load". When a maximum number of deadline misses or maximum laxity is specified for a system based on its working environment, these quantitative data are used in considering the suitability of the system for a given application.

In DIN 66273 timeliness is evaluated in a similar fashion and expressed as a quantitative (benchmark) criterion. Based on the performance indicators gathered, three benchmark criteria are defined, with reference values correlating to values obtained from other implementations, by simulation, or as estimates based on expert knowledge, and B representing the actual load and E the work effect (throughput):

L1 – performance $L1 = \frac{B}{B_{ref}}$ (B_{ref} representing the reference load vector)

L2 – processing time $L2 = \frac{t}{T} t_{ref}$ (t_{ref} representing the reference processing time value)

L3 – timeliness $L3 = \frac{E}{B}$ $(L3 \leq 1)$

In the following sections several aspects of timeliness as a quantitative (benchmark) criterion are presented.

Execution Time Benchmarks (Q0.1)

The *Rhealstone* benchmark is one of the first representatives of *synthetic benchmarks*. Its test series are applied directly on the test platform, and the measured results provide the benchmark reference value to be compared with other systems' reference values. The test suite is meant to check platforms for determinism. Other representatives of this group are: *SNU Real-time Benchmarks*, *PapaBench* and *RT_STAP*; *database and telecommunication benchmarks* and *MiBench* also belong to this group, although they define test program suites for specific applications.

The *Hartstone* benchmark is one of the first representatives of *application-oriented benchmarks*. The test series are built from application tasks in the form of different patterns, which represent operation scenarios considered as typical. They are meant to check platforms for timeliness. Other representatives of this group are the POSIX benchmarks, which were mainly designed to check the timeliness of test platforms, but can also be used to measure the timeliness of applications.

If a specific application cannot be structured in concordance with any of the application-oriented benchmarks, it is more suitable to use a monitor or a timing/schedulability analyser to measure response times and latencies under different application scenarios. Once again, it is important that specifications include all relevant scenarios with acceptable response time and latency values.

An improved (grey-box) variant to determine mere timely deterministic behaviour is deadlock-free execution (deadlock prevention). This requires some knowledge about the system architecture and/or the structure of the application program, which needs to be considered or structurally enhanced by mechanisms for deadlock prevention similar to the ones dealt with in the section on fault prevention.

Response Times to Events Occurring (Q0.2)

Response times relate to activities with time-frames that have to be obeyed for correct functioning. The activities are required to finish on time, and this is sufficient to call them real time. Characteristic properties of typical embedded (real-time) systems, but also of real-time operating systems are the following:

Interrupt latency: duration from the occurrence of an interrupt to the start of an application-specific service task

System call service time: the maximum service times must be bounded and independent on the number of (active) objects in a system

Mask time: the maximum time the operating system and drivers mask interrupts must be bounded and as short as possible

Times to Detect and Correct Errors (Q0.3)

Times to detect and correct errors are related to activities that are responsible for, e.g. data transfers or signal filtering. These times represent the corresponding activities' *response times* that have to be bounded, since overruns can lead to deadline misses or malfunctions within a system.

MTBF, MTTF and MTTR (Q0.4)

Mean Time Between Failures (MTBF), Mean Time To Failure (MTTF) and Mean Time To Recover (MTTR) are benchmarks that are obtained during system operation and can be evaluated statistically. Since they represent mean values, they are only applicable to soft real-time systems. Also these values are of limited use, since they offer no guarantees on the time-frames they state.

5.4.2 Noise Suppression (Q1)

In signal filtering and recognition operations, noise has to be kept within certain limits. Hence, also the time to adapt to interferences, representing very low/high signal/noise (S/N) values should be limited. The S/N values and times specified are regarded as bounds of functional correctness (X1) and (response) time-frames of the corresponding activities (Q1), respectively.

5.4.3 Capacity Reserves (Q2)

Benchmarks, particularly application-oriented ones like Hartstone, provide performance indicators to assess timeliness. These measurements are helpful to determine capacity reserves in the (response) time-frames of activity executions, which relate to their specified *response times*. Regarding physical components, capacity reserves pertain to free slots for hardware board interfaces, differences between required and available memory capacity, maximum number of active tasks, etc.

5.4.4 Overall Project Costs (Q3)

The "bottom line" of all considerations is often cost, and a general optimisation objective is to minimise it. As overall cost comes about in an additive way, there is wide room to consider trade-offs and to make compromises in choosing hardware and software components under observation of the limitations specified.

6

Design of Real-time Systems for QoS

It was discovered early, that in order to achieve the desired QoS properties a design methodology for real-time systems has to include appropriate measures to ensure that the QoS criteria are considered during the entire life-cycle. These have been joined in the framework of the ISO/IEC 13236 [31] and ISO/IEC TR 13243 [32] standards for QoS in information technology, and the standard IEC 61508 [28], which includes the necessary activities for safety-related systems from a project's start until the end of its life-cycle. Like safety, security is also an issue that is gaining importance for embedded (real-time) applications [76], and which must be dealt with during the design phase, too.

In the following sections the mentioned QoS criteria are addressed from different phases and aspects of real-time system design, where they should be considered in order for a system to fulfil these requirements. Apart from the general ISO/IEC 13236 and ISO/IEC TR 13243 standards, which mainly deal with system functionality and development process, more specific QoS standards for real-time systems are addressed here. Timeliness, being the most important QoS property of real-time systems, is considered in detail in the next chapter. It is, however, affected by the issues addressed here and cannot, therefore, be maintained in the long run unless they are also satisfied.

6.1 Design for Predictability and Dependability

6.1.1 Design for Predictability

For real-time systems the foremost property is predictability. It also pertains to behavioural predictability, which is addressed in the following sections. However, in the real-time domain usually a system's timeliness is meant, which represents the property of the system whether all its actions can be performed in time during its entire up-time. Such a system is considered to behave in a (temporally) predictable way, and is said to "operate in real time".

During design, predictability is supported by carefully planning the order of activities as well as taking care of their durations. Often the activities are executed periodically and, hence, a main loop, named *cyclic executive*, is built around them. It invokes them in a certain order when a timer, to which their period is assigned, times out. Naturally, the sum of their durations shall not exceed this period for them to finish in time. These activities, usually called *tasks*, do not necessarily have a unique and fixed period. Hence, dynamic scheduling algorithms have been devised (e.g. the Rate Monotonic one (RM)), which dynamically assign priorities to tasks based on their relative urgencies (reciprocal values of their periods). Also, some or all of the tasks may not be periodic, but still have temporal constraints on their execution. For these cases other non-priority- (time) oriented scheduling algorithms have been devised (e.g. Earliest Deadline First (EDF), or Least Laxity First (LLF)). They are considered in a special kind of performance analysis, named *schedulability analysis*, which determines the temporal predictability of diverse execution scenarios to discover bottlenecks and, foremost, to foresee the temporal predictability of a system's activities.

There are two kinds of design approaches that enable reasoning on the timeliness of task executions — formal and non-formal. The formal design methods encompass temporal state automata (e.g. Communicating Shared Resources (CSR) [69], timed Petri nets [79] or UML state charts). The non-formal design methods encompass Gantt diagrams and derivatives thereof (e.g. UML sequence diagrams and UML timing diagrams). While formal design methods can deliver estimates of an activities' execution time, it is still up to the scheduling algorithm to ensure their timely execution in correspondence with their periods/deadlines.

6.1.2 Design for Availability

Permanent readiness in high-availability systems requires that the systems be designed for non-stop operation. Of course, such systems also require maintenance and occasional software upgrades. For this purpose, re-configuration management mechanisms have been developed. Initially, static re-configuration has been used, representing strictly defined operation scenarios (e.g. manufacturing lines, space shuttle or avionics). With the advent of reactive systems the need for dynamic re-configuration arose, where the lines among phases are not so strict, and where there may be scenario parts that are interchanged leaving the rest intact and functioning.

Re-configuration management should support implementation platforms with the following attributes:

Heterogeneity: optimising architectures for performance, size and power consumption requires the most appropriate implementation techniques to be used; implementations require software-configurable hardware

(Hard) real time: usually there are significant real-time constraints on embedded systems

Re-configurability: an execution environment must allow hardware and software resources to be re-allocated dynamically

Execution environments supporting dynamic re-configuration encompass the following features:

- Specification of hardware and software configurations with well defined re-configuration scenarios — conditions and methods for re-allocation of hardware/software components and their interconnections
- Monitoring of local state changes and delegation of state changes to affected processing nodes during re-configuration
- Minimum and predictable overhead to overall execution time through short and well defined re-configuration actions

During dynamic re-configuration, application data must remain consistent and real-time constraints must be satisfied. To fulfil the mentioned requirements, these issues must be addressed at multiple levels:

Hardware level: At the lowest level, the hardware must be re-configurable. Software-programmable hardware components have best inherent hardware support, since their functions can be changed by memory contents. Also, internal hardware structures can be designed in a way restricting dangerous conditions that may damage hardware.

Middleware level: At a slightly higher level, the internal state of the system must be managed under changing tasking. Operating systems have evolved to support flexible implementations of multiple tasks on a single processor in the form of time-sharing and/or multitasking. Typically, however, this alone is not enough for low-level efficiency and/or hard real-time conditions.

Software level: To represent the behaviour of an embedded real-time system, generally three viewpoints must be considered: (1) the external functional viewpoint, representing the operation scenarios, (2) the internal functional viewpoint, which models the dynamical state changes of the system, and (3) the specification of the hardware and software architectures together with the mapping of software onto hardware components. These are the data required by the re-configuration management execution environment.

Re-configuration management for embedded (real-time) systems has mostly been dealt with in connection with their hardware/software co-design [47, 63, 78]. Besides defining diverse (dynamic) operation scenarios, two main goals have been achieved by this approach: (1) achieving fault tolerance by system design [25, 40], and (2) fast scenario switching for industrial automation and telecommunication systems [19, 27, 39, 63].

6.1.3 Design for Safety

In the late 1980s, the International Electrotechnical Commission (IEC) started the standardisation of safety issues in computer control. Four Safety Integrity Levels (SIL) were defined, with SIL4 being the most critical one. Activities were prescribed at different levels and phases of system development (e.g. coding standards, dynamic analysis and testing, black-box testing, failure analysis, modelling, performance testing, formal methods, static analysis, modular approach), which are desired or mandatory, and approaches that are allowed or required in order to fulfil the requirements of a certain Safety Integrity Level. These rules form the standard IEC 61508 for the life-cycle management of instrumented protection systems. As can be seen from Fig. 6.1, the safety life-cycle encompasses the entire production cycle from a system's design to its decommissioning.

SIL Flowchart

The flowchart in Fig. 6.1 represents the safety life-cycle of an Equipment Under Control (EUC) as a whole in its entirety. Such EUC is composed of one or more Electrical/Electronic/Programmable Electronic (E/E/PE) devices, which as a system have to fulfil individual as well as collective safety requirements. The single steps of the safety life-cycle for an EUC have the following meanings:

1. Concept: all information on the EUC (physical, legislative, etc.), relevant to the following steps need to be assembled
2. Overall scope definition: specification of the hazards and risks (e.g. process hazards, environmental hazards)
3. Hazard and risk analysis: is performed to determine the hazards and hazardous events or sequences of events of the EUC (and the EUC control system) for all foreseeable cases including fault conditions and misuse and to determine the associated risks
4. Overall safety requirements: a specification of the overall safety requirements is developed in terms of the safety function requirements and the safety integrity requirements in order to achieve the functional safety required
5. Safety requirements allocation: places the safety functions from the specification produced previously on the designated E/E/PE or on other-technology safety-related systems as well as external risk reduction facilities, and allocates a safety integrity level to each safety function
6. Overall operation and maintenance planning: is carried out for the E/E/PE safety-related systems to ensure that the required functional safety is maintained during operation and maintenance
7. Overall safety validation planning: facilitates the overall safety validation of the E/E/PE safety-related systems
8. Overall installation and commissioning planning: is carried out to ensure that during these phases the required functional safety of the E/E/PE safety-related systems is achieved
9. Safety-related systems (E/E/PES) realisation: comprises the creation phase of the E/E/PE safety-related systems where the specified functional safety and safety integrity requirements have to be obeyed
10. Safety-related systems (other technology) realisation: comprises the creation phase of other safety-related systems where the specified functional safety and safety integrity requirements for these specific technologies have to be obeyed (outside the scope of this standard)

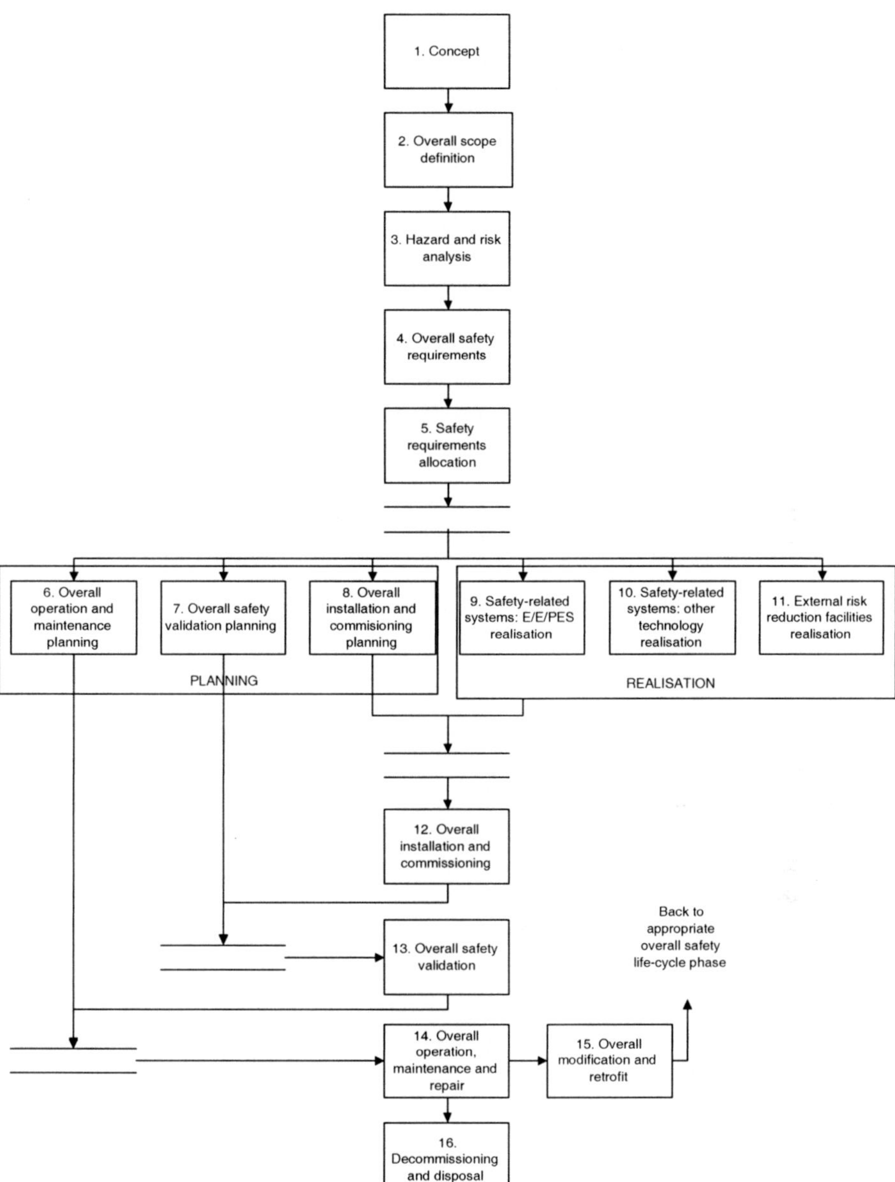

Fig. 6.1. Safety life-cycle according to IEC 61508

11. External risk reduction facilities realisation: comprises the creation of external risk reduction facilities to meet the specified safety functions and the safety integrity requirements thereof (outside the scope of this standard)
12. Overall installation and commissioning: comprises the installation/commissioning phase of the E/E/PE safety-related systems
13. Overall safety validation: is to validate that the specified functional safety and safety integrity requirements for the E/E/PE safety-related systems are met
14. Overall operation, maintenance and repair: comprises the with respect to safety functionally intact operation, maintenance and repair of the E/E/PE safety-related systems
15. Overall modification and retrofit: is meant to ensure that the functional safety for the E/E/PE safety-related systems is appropriate during and after these phases
16. Decommissioning or disposal: is meant to ensure that the functional safety of the E/E/PE safety-related systems is appropriate during and after these phases on the EUC and its control system

Safety Integrity Levels

The standard IEC 61508 details the requirements necessary for a system to qualify for each of the four Safety Integrity Levels. These requirements are more rigorous at higher levels of safety integrity in order to achieve the required lower likelihood of dangerous failures.

An E/E/PE safety-related system usually implements more than one safety function. If the safety integrity requirements for these safety functions differ, unless there is sufficient independence between their respective implementations, the requirements applicable to the highest relevant Safety Integrity Level shall apply to the entire E/E/PE safety-related system. If a single E/E/PE system is capable of providing all required safety functions, and the required safety integrity is less than that specified for SIL1, then IEC 61508 does not apply.

For safety-related systems two kinds of requirements are defined:

1. *Safety function requirements:* the safety functions that have to be performed
2. *Safety integrity requirements:* the reliability with which the safety functions have to be performed

By *functional safety* the ability of a safety-related system to carry out the actions necessary to achieve a safe state for the equipment under control or to maintain a safe state for the equipment under control is meant and relates to *safety*. *Safety integrity* is the likelihood of a safety-related system to achieve safety functions required under all conditions stated within a given period of time; it relates to *reliability*.

The Safety Integrity Levels are defined in Table 6.1 by the probabilities of "failure on demand (FOD)", "protective system technology", and "protective system design/testing/maintenance (D/T/M) requirements". We can observe how the increasing *complexity* of designs and components used also increases the cost and maintenance interval rate of systems. With falling maximum failure probability also the need to allow for common-cause failures is raised, which adds to complexity and cost. On the other hand, *simplicity* in design and usage of standard components with lower sensing and actuation (S&A) diversity prolong the required maintenance intervals and lower the need to allow for common-cause failures. Failure probability is the direct opposite of *dependability*.

Table 6.1. Safety Integrity Levels

FOD probability	
SIL1	0.1–0.01
SIL2	0.01–0.001
SIL3	0.001–0.0001
SIL4	0.0001–0.00001
System technology	
SIL1	Standard components, single channel
SIL2	Predominantly standard components
SIL3	Multiple channel with S&A diversity
SIL4	Specialised designs
D/T/M requirements	
SIL1	Relatively inexpensive
SIL2	Moderately expensive
SIL3	Expensive
SIL4	Very expensive
Test interval	
SIL1	> 3 months
SIL2	< 3 months
SIL3	1 month
SIL4	< 1 month

Apart from the above-mentioned process techniques to achieve system safety, some design techniques have also been devised. The mentioned design techniques, representing vital parts of a system's development phase, form parts 6 and 9 of the safety life-cycle in Fig. 6.1. Some of them are summarised in Table 6.2 together with their importance to the individual *Safety Integrity Levels*.

We can observe similarities in recommended design techniques among SIL1 and SIL2, and among SIL3 and SIL4. The main differences are in the rigour of their application. There is a noticeable rise in the required effort and rigour of the methods to be applied between SIL2 and SIL3 (e.g. the use of formal methods).

In the following sections some of the mentioned design techniques are discussed and some further ones, which have been devised especially for fault removal or fault tolerance in dependable real-time systems, are presented.

Table 6.2. Software practices from IEC 61508-3 by category

Practice	61508-3	SIL1	SIL2	SIL3	SIL4
Coding standards					
Use of coding standard	B.1	HR	HR	HR	HR
No dynamically allocated variables	B.1	—	R	HR	HR
Dynamic analysis and testing					
Test case execution from cause consequence diagrams	B.2	—	—	R	R
Structure-based testing	B.2	R	R	HR	HR
Black-box testing					
Equivalence classes and input partition testing	B.3	R	HR	HR	HR
Failure analysis					
Failure modes, effects and criticality analysis	B.4	R	R	HR	HR
Formal methods modelling	B.5	—	R	R	HR
Performance modelling	B.5	R	HR	HR	HR
Timed Petri nets	B.5	—	R	HR	HR
Performance testing					
Avalanche/stress testing	B.6	R	R	HR	HR
Response timings and memory constraints	B.6	HR	HR	HR	HR
Performance requirements	B.6	HR	HR	HR	HR
Semi-formal methods					
Sequence diagrams	B.7	R	R	HR	HR
Finite state machines/state transition diagrams	B.7	R	R	HR	HR
Decision/truth tables	B.7	R	R	HR	HR
Static analysis					
Boundary value analysis	B.8	R	R	HR	HR
Control flow analysis	B.8	R	HR	HR	HR
Fagan inspections	B.8	—	R	R	HR
Symbolic execution	B.8	R	R	HR	HR
Walk-throughs/design reviews	B.8	HR	HR	HR	HR
Modular approach					
Software module size limit	B.9	HR	HR	HR	HR
Information hiding/encapsulation	B.9	R	HR	HR	HR
Fully defined interface	B.9	HR	HR	HR	HR
Total recommended (R)		12	12	3	1
Total highly recommended (HR)		6	10	20	22

Legend: HR highly recommended; R recommended; — no recommendation

6.1.4 Design for Reliability

Design for Testability

Maintainability and testability concern systems on which it is possible to act in order (1) to avoid the introduction of faults and, (2) when faults occur, to detect, localise and correct them. The means to tackle these issues can be divided into the following categories [21]:

Fault prevention aims to reduce the creation and occurrence of faults during the life-cycle of a system. It includes the measures to ensure that all *applicable physical constraints are met* and that *only static and real features are used*. As a precaution, to prevent faults in the design phase as well as to ensure maintainability, the concern for bounded complexity of the design as a whole and of its parts has to be taken into account as well. This eases portability and flexibility of a solution.

Fault removal aims to detect and eliminate existing faults. It is addressed by verification and validation of a system against its specifications.

Fault tolerance aims to guarantee the services provided by a system despite the presence or occurrence of faults. This issue is meant to ensure permanent readiness, simultaneous operation, predictability, robustness, graceful degradation and prevention of deadlocks.

Fault forecasting aims to estimate the presence of faults (number and severity).

Apart from the suppression of faults in real-time systems, the equally important timeliness issue has to be considered. In real-time systems, timing faults and functional faults are treated equally. Since these faults cannot be removed, they can only be addressed by fault forecasting and prevention, and handled by fault tolerance (e.g. graceful degradation) mechanisms.

To analyse the timing behaviour in a system, hardware and software monitors can be applied, and timing analysers used on the compiled code to estimate worst-case response times, error detection and correction times and signal-to-noise ratio bounds in signal processing applications. The quantities TBF, MTTF and MTTR are hard to measure, and exhaustive systematic testing has to be applied on a final system, possibly permanently damaging or destroying it.

Fault Prevention

Once the elicitation of a client's needs has been completed, a system is to be defined that fulfils these needs. This work leads to the expression of *specifications*. Since they are the first source of faults, they have to be well pondered and precisely formulated. To reduce the possibility of faults at this phase in the design process, the specifications may be checked for conformity with the following criteria:

1. *Semantic criteria*
   Non-ambiguity: each element of the specification should have one interpretation only
   Completeness: all aspects of operation should be covered and boundary conditions defined
   Consistency: there should be no conflicts between the elements of the specification.
   Traceability: the customer needs and specification elements have to be clearly correlated
2. *Syntactic criteria*
   Concision: unnecessary and misleading terms should be avoided
   Clarity: the sentences of the specification should be easy to read, i.e. short and simply formed; the text flow should be straightforward, i.e. unnecessary references and jumps should be avoided
   Simplicity: the concepts addressed have to be simple, their number has to be limited and they should be loosely coupled
   Comprehension: reading of the text has to facilitate the understanding of the semantics

Non-conformity to these criteria increases the risk of faults. These criteria also apply to design models. Methods to verify specifications comprise reviewing, scenarios and prototyping.

Review processes are carried out by humans analysing the contents of specifications. There are two general approaches to reviews: walk-through and inspection. A reviewer may search for faults or risks of faults in a specification model. More reliable results are produced if a review is conducted by a person or team that was not involved in the specification. Resulting notes on the (potential) faults are returned to the specification's producer.

From *scenarios* input/output sequences that simulate the interaction between the system specified and its environment are derived. They are presented to the client, who approves or disapproves them.

The result of *prototyping* is a tool, built from the specification document, which simulates the system's interactions with its environment. Its use by the client allows the detection of errors or misconceptions in the specification. Herewith, also possible sources of timing faults can be detected and dealt with appropriately.

The same guiding principles as for specifications remain pertinent also when choosing an appropriate design model. Methods of adequate expressiveness have to be selected in concordance with the characteristics of the system designed. The more the concepts to be modelled concur with the model features, the *simpler* the solution will be, and the lesser will be the probability of introducing faults. Versatile design methods for the hardware and software constituents of systems have already been partly unified (e.g. in the UML methodology dealt with in detail in the next chapter). There are, however, still aspects of hardware (e.g. boards) and software (e.g. temporal behaviour and analysis) design that are only partly addressed by the more general design methods. Hence, the pertaining properties have to be specified as well as possible in the framework of a system and the detailed design has to be checked separately to fulfil their specifications in the "bigger picture".

The choice of an adequate design modelling tool is not sufficient to design a faultless system. It is equally important to employ a proper design process. Indeed, faults can also be due to the designers' inability of deducing a correct model due to the expressive means available or due to an inappropriate choice of design phases. One of the prominent design processes for faultless systems is the standard SIL-based process model presented in the next section.

Example: Coding Rules for Safety-critical Applications

Most serious software development projects use coding guidelines. In connection with a consistent design method and process, they are very important to ensure coherent software artifacts in a larger and/or evolving project team. They define the ground rules for the software to be written "by hand": how it should be structured, which language features should be used and how. According to [26] there is, surprisingly, little consensus on what a good coding standard is. In this article, ten rules were defined with special emphasis on safety-critical applications:

1. Restrict all code to very simple control flow constructs — do not use *goto* statements, *setjmp* or *longjmp* constructs, or direct or indirect recursion.
2. Give all loops a fixed upper bound. It must be at least trivially possible to statically prove that a loop cannot exceed a pre-set upper bound for the number of iterations.
3. Do not use dynamic memory allocation after initialisation.
4. No function should be longer than what can be printed on a single sheet of paper in a standard format with one line per statement and one line per declaration (typically this means a maximum of 60 lines of code per function).
5. The code's assertion density should average to a minimum of two assertions per function. Assertions must be used to check for anomalous conditions that should never happen in real-life executions. They must be free of side effects and should be defined as Boolean tests. When an assertion fails, an explicit recovery action must be taken, such as returning an error condition to the caller of the function that executes the failing assertion. Any assertion for which a static checking tool can prove that it can never fail or never hold violates this rule.
6. Declare all data objects at the smallest possible level of scope.
7. Each calling function must check the return value of non-*void* functions, and each called function must check the validity of all parameters provided by the caller.
8. The use of a preprocessor must be limited to the inclusion of header files and simple macro definitions. Token pasting, variable argument lists (ellipses) and recursive macro calls are not allowed. All macros must expand into complete syntactic units. The use of conditional compilation directives should be kept to a minimum.

9. The use of pointers must be restricted. Specifically, no more than one level of de-referencing should be used. Pointer de-reference operations may not be hidden in macro definitions or inside *typedef* declarations. Function pointers are not permitted.
10. All code must be compiled from the first day of development, with all compiler warnings enabled at the most pedantic setting available. All code must compile without warnings. All code must also be checked daily with at least one, but preferably more than one, strong static source code analyser and should pass all analyses with zero warnings.

Of course the choice of a programming language is a key consideration in this concern. Since for programming aerospace, industrial automation and automotive applications mostly C is used, in this case it was also the language chosen and the rules apply to it. The benefit of using a smaller number of coding rules lies in the fact that it is more likely that they are going to be used as opposed to hundreds of rules and guidelines. The author of [26] claims, that the quality of source code improved within the NASA/UPL Laboratory for Reliable Software by using these rules.

While the first few rules ensure the creation of clear and transparent control flow structures that are easier to build, test and analyse, the other rules address some "fancy" features of the language C, which work in principle, but are quite error-prone. Dynamic memory allocation is discouraged because of the run-time problems with temporally deterministic memory management in environments with potentially very limited memory resources. Some rules address the fault detection and removal techniques described later. These rules correlate very strongly with the guidelines and regulations that have been assembled and classified in the standard IEC 61508 described previously.

Fault Removal

The fault prevention techniques presented so far are useful during specification and design. Here we examine a designed system (model) in order to detect the possible presence of residual faults. Some techniques require the presence of specifications, while others are derived from the designed system. Their demonstration strength ranges from *partial functional simulation* to *complete formal proof*. They may apply to a system that has not yet been fully designed or implemented (in hardware and software) or to a final product.

Verification without specifications aims at checking the design model (system) for the existence of undesired properties such as deadlocks or non-determinism. Verification with specifications is possible by the following methods:

- Bottom-up reverse transformation of the designed system/model into a specification and its comparison with the actual specification
- Top-down verification by bivalent specification to designed system/models and their comparison by simulation sequences or by checking functional static and/or dynamic properties

A simulation sequence, also known as "test sequence", generally consists of tuples of inputs and expected outputs, which are applied to the design model (system) and compared with the results obtained. If they are equal to the expected outputs, the system "conforms to the specifications". For real-time systems, they should be enhanced by the time dimension. The level of trust in the results depends on the quality of the test sequence. Generally, it depends on the level of structural knowledge about the design model (system) — we speak about black-box (no knowledge), grey-box (little knowledge) and white-box (complete knowledge) tests.

Short Test Sequences

A system may be designed in many different ways depending on the nature of the problem domain, required functionality, and a number of QoS parameters resulting in very different implementations: circuits containing electronic and mechanical components, logic gates (arrays), or sequential program code and very likely combinations thereof. Specific design methods naturally integrate the previously stated remarks on testability connecting the most suitable test methods with the respective design methods.

Often ideas on testability can be transferred from mechanical to electrical engineering and further to program design and test, which makes integration tests easier with hybrid systems being composed of all three and possibly additional (e.g. chemical) kinds of devices.

The subsequent examples show how fault detection and removal techniques have been integrated with specific design methods. This, however, does not mean that the use of the principles presented is restricted to the domains indicated.

Built-In Test (BIT): This technique consists of adding a specific standard interface to the system under test, which controls and facilitates access from the external tester, thus increasing controllability and observability (see Fig. 6.2). As a consequence, the test sequences are simpler, and their application is facilitated. Hence, testing is also simpler. The drawback is that the technique introduces some additional overhead into hardware components and/or software code.

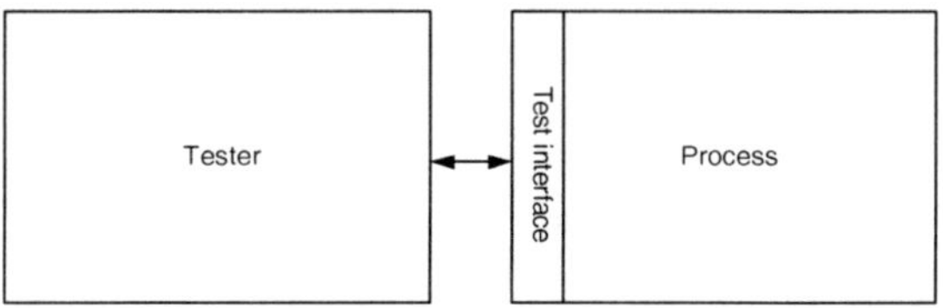

Fig. 6.2. BIT test schema

Boundary Scan: In IEEE 1194-1 the IEEE Joint Test Action Group (JTAG) defined a standardised interface to test integrated circuits. Today, the majority of integrated circuit manufacturers uses this standard to test their designs of application-specific integrated circuits, microprocessors and microcontrollers.

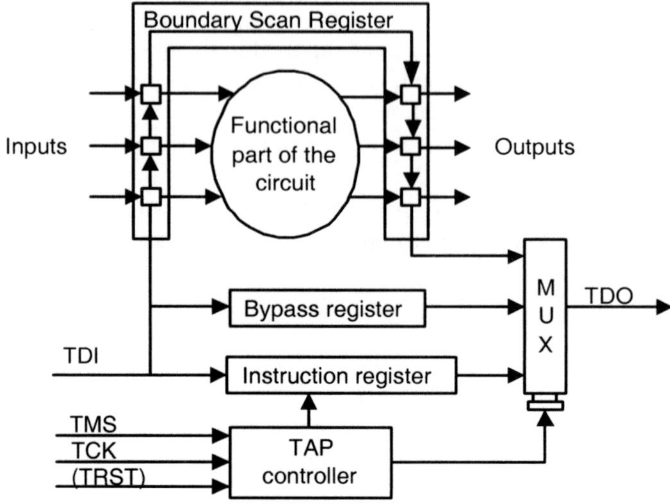

Fig. 6.3. Schema of Boundary Scan Register

addition of some test inputs/outputs and an internal logic:

Test bus ("Test Access Port" (TAP)) composed of a series of input/output signals (TDI, TDO) and a test clock (TCK)

Integrated logic module comprising:

1. A series register ("Boundary Scan Register") to read in the test inputs and read out the test results
2. A bypass register to reach other modules located below in the test chain of the system's modules
3. An automaton (TAP Controller) associated with an instruction register, which may process certain test operations. It controls the various operations in normal or test mode. In normal mode, the boundary scan register cells are set so that they do not affect the circuit. In test mode, however, the cells are latched in a way enabling a data stream between them — once a complete data word has been read in, it can be latched into place. The cell at the destination pin of the circuit can be programmed to verify if the circuit trace correctly interconnects the two pins (input and output).

Each cell (see Fig. 6.4) of the *Boundary Scan Register* has four modes of functioning. In "normal mode" (1) the cell's output multiplexer (*Output MUX*) transfers the data coming from an input pin of the circuit to the output of the cell (*Data Out*), which is connected to the input of the core logic; in "update mode" (2) the output multiplexer sends the contents of the *Parallel Output Register* to the cell's output; in "capture mode" (3) the input data (*Data In*) is directed by the input multiplexer (*Input MUX*) to the *Shift Register* in order to be loaded when the *Clock DR* occurs; in "shift mode" (4) the bit of each cell of the *Boundary Scan Register* (its *Shift Register*) on the *Inputs* side is sent to the following cell via the *Scan Out* line, whilst the signal *Scan In* coming from the previous cell is loaded into the flip-flop of the cell's *Shift Register*. The output cells use the same principle. As the cells can be used to force data into the circuit, they can set test conditions. The relevant states can then be fed back into the test system by clocking the data word back so that it can be analysed.

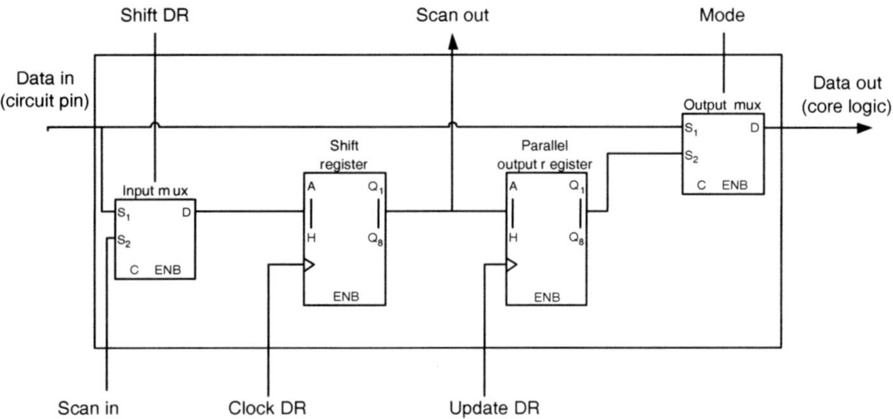

Fig. 6.4. Boundary Input Cell

The circuit logic described allows one to control all inputs and outputs of the circuit tested with the help of a shift register, which can be loaded and unloaded in series. It is also possible to pass information across a circuit in order to reach a circuit situated below it, or to receive information coming from this circuit. Adopting this technique, it is possible for a test system to gain test access to a circuit board. As most current boards are very densely packed with components and lines, it can be very difficult for test systems to access the relevant areas of a board via probes for test purposes. The boundary scan makes this possible. In some cases the boundary scan technique is not applied to a circuit as a whole. Then, we say that this circuit implements a "partial scan" as opposed to a "full scan".

The same hardware circuit test logic can also be applied to software. "Scan design" is appropriate for circuits or software modules having a few inputs and outputs, since otherwise too long test sequences would result. A natural approach to solve this problem is to partition circuits/software into parts called "scan domains". Once the partial "unit tests" are complete, "integration tests" can be performed.

Built-In Self Test (BIST): The BIT technique (see Fig. 6.2) requires an external tester (e.g. "off-chip test" in hardware design), while with the BIST technique (see Fig. 6.5) the tester is integrated into the tested system (e.g. "on-chip test" in hardware design). The method requires a greater investment in design. The BIST approach improves the principle of integrating the tester functions into the system under test. This solution is only acceptable if its complexity and the expenditure are not excessive. It is justified when the coverage of faults tested is increased considerably. The technique allows the system to test itself, usually off-line, with the normal function of the system being suspended during the test. Usually, the test operations are run during power-up or maintenance operations. Some fault tolerance techniques require on-line monitoring of inputs and outputs. In this case, BIST functions are integrated into the normal functionality of the system under test. BIST techniques are integrated in more and more industrial products.

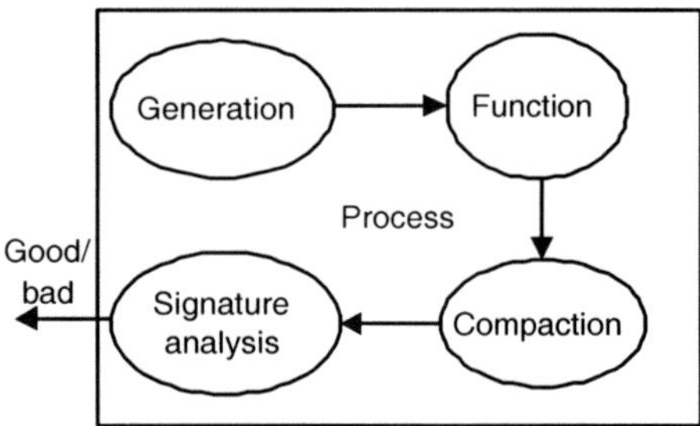

Fig. 6.5. The BIST test schema

As an example, the generation of pseudo-random test patterns for BIST employing a Linear Feedback Shift Register (LFSR) is presented. The 3-bit LFSR shown in Fig. 6.6 is a synchronous sequential circuit, using D flip-flops and XOR gates, which generates a pseudo-random output pattern of 0s and 1s. Suppose the circuit is initialised in state (Q1,Q2,Q3)=(1,1,1); it produces a cyclic output sequence with an (Q1,Q2,Q3) output vector for any clock pulse. Such an LFSR can be used as a test sequence generator in an integrated circuit (DUT – Device Under Test) as shown in Fig. 6.7. It can also be used as a compaction circuit (PSA – Parallel Signal Analyser) in order to reduce the length of the output sequence (see Fig. 6.8).

Fault Tolerance

Fail-safe or *fail-secure* is the property of a system/device that, if (or when) it fails, it fails in such a way not to cause harm, or only minimum harm, to other devices or danger to personnel.

Fail-safe operation means automatic protection of programs and/or processing systems when a hardware or software failure is detected in a computer system.

Fault tolerance is the property that enables a system to continue operating properly upon occurrence of a failure in (or of one or more faults within) some of its components.

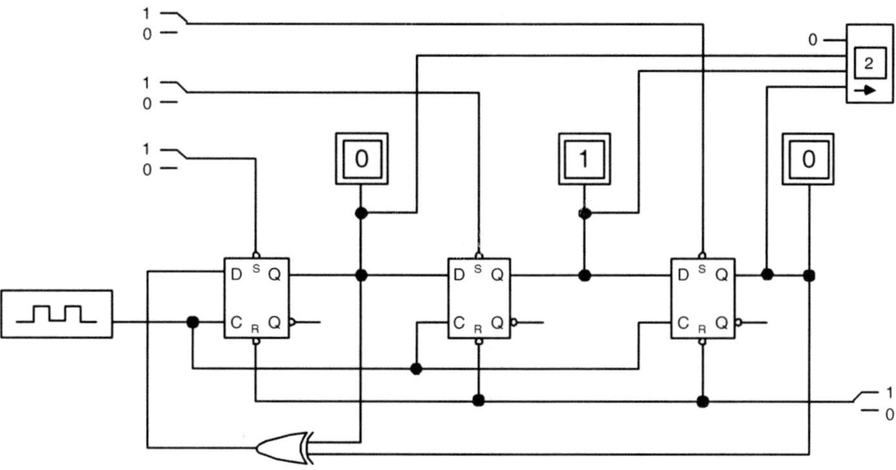

Fig. 6.6. 3-bit LFSR

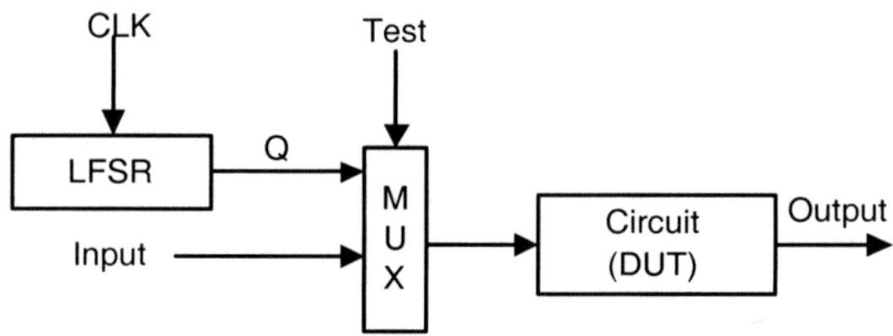

Fig. 6.7. Test sequence generation

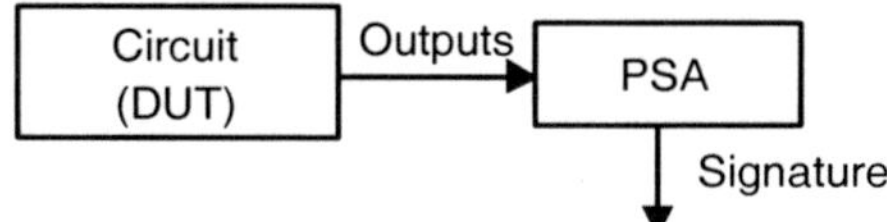

Fig. 6.8. Test result compaction

Protective *fault tolerance mechanisms* are defined and implemented during the system creation phases of the life-cycle, whereas their actions are effective during the operational phase. Fault prevention and removal techniques allow one to increase reliability (reducing the probability of fault occurrences), or availability in the case of repairable systems (e.g. detect-and-repair mechanisms). Observing safety criteria leads one to examine techniques related to *on-line testing* and *fail-safe design*. Here, we want to provide a system designed with the highest degree of dependability by integrating mechanisms that allow for full continuation of its mission despite the presence of faults, thus increasing its integrity. Such mechanisms may deliver full service at no reduction in performance. Due to the presence of faults, however, the performance usually decreases to an acceptable level ("graceful degradation"). Employing fault tolerance techniques has a direct positive impact on *safety*, *reliability*, and *availability*.

In the sequel, the basic characteristics of fault tolerance are dealt with in more detail. Fault-tolerant systems are also characterised in terms of planned and unplanned service outages. They are usually regarded/measured from the application point of view, and quantified by *availability* as a percentage (e.g. an availability of 95.9% means that a system is up this fraction of time).

Fault tolerance by replication: The first fundamental characteristic of fault tolerance is "no single point of failure". It is addressed by spare components (replicas) in three ways:

1. *Replication:* providing multiple identical instances of the same system or device, directing tasks or requests to all of them in parallel and choosing the correct result on the basis of a quorum
2. *Redundancy:* providing multiple identical instances of the same system or device and switching to a remaining one in case of a failure (failover)
3. *Diversity:* providing multiple different implementations of the same specification, and using them like replicated systems to cope with errors in a specific implementation

A good example of replication is a redundant array of independent disks (RAID) representing a fault-tolerant storage device using data redundancy.

A *lockstep fault-tolerant system* consists of replicated elements operating in parallel. At any time, all replicas of each element should be in the same state. The same inputs are provided to each replica, and the same outputs are expected. The replicas' outputs are compared using a voting circuit. A system with two replicas for each element is termed *Dual Modular Redundant* (DMR). In this case the voting circuit can detect a mismatch, only, and recovery relies on other methods. A system with three replicas for each element is termed *Triple Modular Redundant* (TMR). The voting circuit can then determine which replication is in error and output the correct result based on a two-to-one vote. After that, the voting circuit usually switches to DMR mode. This model can be applied to a larger number of replicas, e.g. *pair-and-spare*. Here, two replicated elements operate in lockstep as a pair, with a voting circuit that detects any mismatch between their operations and outputs a signal indicating when there was an error. Another pair operates exactly in the same way. A final circuit selects the output of the pair not being in error.

No single point of repair: With this option enabled, a system must continue to operate without interruption during the repair process if it experiences a failure.

Fault isolation to the failing component: When a failure occurs, a system must be able to isolate the failure to the erroneous component. This requires dedicated failure detection mechanisms to be added, solely for the purpose of fault isolation.

Fault containment: This is to prevent the propagation of failures. With some failure handling mechanisms a failure may be propagated to the rest of a system, possibly causing it to fail completely. Mechanisms that isolate a failing component to protect the system are required.

Reversion modes: With some failure mechanisms the survivability of a system, its operator, or the final results may be endangered. To prevent such losses, mission-critical systems (e.g. in defence or avionics) provide for safe modes using *interlock* mechanisms or software.

Example: Safety Shell

Since the two most important properties of real-time systems are dependability and timeliness, for their safety-oriented design a so-called system "safety shell" has been defined [45, 46]. It combines a suitable combination of the above listed options and mechanisms (replication, no single point of failure, fault containment) to ensure timely, fault-tolerant execution.

The software safety shell shown in Fig. 6.9 depends on a guard to ensure safe execution in a system. The guard is a low-level construct that detects faults and, then, forces the system to transfer to a safe state instead of a hazardous one. As can be seen, the "Primary Control" is guarded from erroneous input from the environment by the "State Guard". Its *functional correctness* is ensured by defining well formed algorithms with strict limitations on their I/O range, and "Exception Handlers" maintaining safe and stable states of operation. At the same time, the timely, synchronous operation of the system is guarded by the "Timing Guard", implemented by, e.g. "watchdogs" for critical operations.

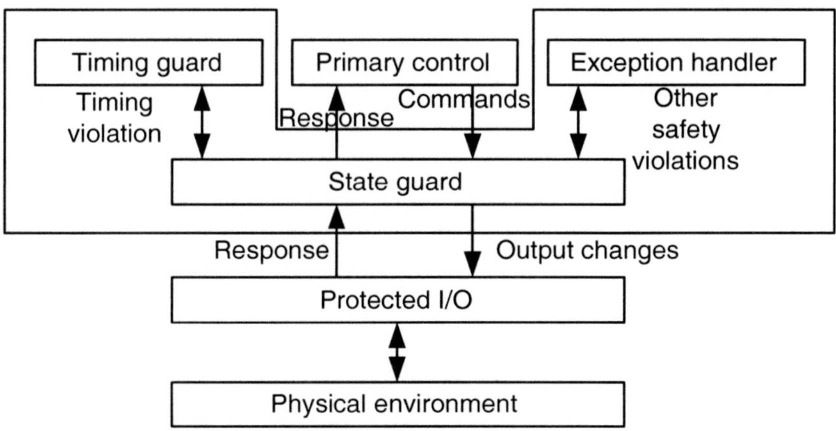

Fig. 6.9. Safety shell

Fault Forecasting

Fault forecasting aims to estimate the presence of faults (number and severity). It implies the means that allow the evaluation or measurement of *reliance*. The assessment approaches available can be categorised into quantitative and qualitative dependability assessment.

The objective of *qualitative assessment* is to examine faults, errors, potentiality of failures, and their effects. The assessment methods are twofold — deductive and inductive. In the deductive approach, the failures anticipated are derived from faults and errors. In the inductive approach, the potential failures are examined from present faults or errors.

In the standard EN 13306 [20] *reliability* is defined as: *"The ability of an item to perform a required function under given conditions for a given time interval,"* where item denotes a device (product) considered for maintenance or maintenance management. *Quantitative assessment* considers reliability as a function of time which expresses the conditional probability that a system has survived in a specified environment until time *t*, given that it was operational at time *0*. This definition is the origin of the quantitative QoS parameters (e.g. MTBF, MTTF and MTTR).

A product's reliability function does not increase with time. Its decreasing tendency is due to the subjacent phenomenon of degradation of electronic devices. In the case of software, once this function has reached a certain level, it remains constant due to the absence of ageing (not considering maintenance operations).

Different types of *reliability tests* are carried out to perform *reliability evaluation* with respect to test stop conditions such as fixed duration, fixed number of faults reached, result-based end criteria, and combinations thereof, and with the possibility of accelerated — *avalanche/stress* — tests.

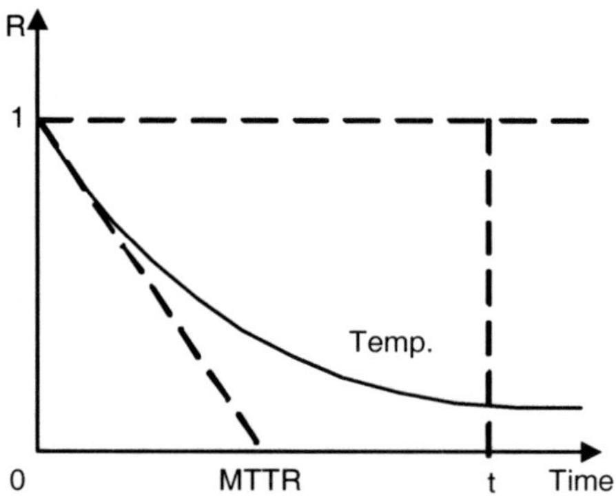

Fig. 6.10. The exponential law

Reliability is analysed by *reliability models*, which are mathematical functions of time. The *exponential law* is the simplest of these laws. It provides the probability of survival by an exponential function, which decreases with time (see Fig. 6.10). The *failure rate* λ expresses the probability of failures occurring per hour, and is generally considered to be a constant throughout time. The $R(t)$ law is often associated with the simple estimators:

MTTF "Mean Time To Failure" (also called Mean Time to First Failure – MTFF) for non-repairable systems (e.g. a mission terminating as soon as a breakdown happens)

MTBF "Mean Time Between Failures" for repairable systems (e.g. a component, which broke down, is repaired/replaced and put back into operation)

If a reliability function described by the exponential law has a constant failure rate, the average value of this function is: MTBF (or MTFF) $= \frac{1}{\lambda}$ and is expressed in powers of 10 hours.

Failure rates are generally estimated from survival tests applied to significant large samples of components. The duration of these tests is short as compared to a product's normal life-cycle. Then, by using acceleration factors, the failure rates of, e.g. electrical circuits are deduced. These tests are thus called *accelerated tests*. A component's degradation process can be accelerated by increasing the temperature, and also by increasing the voltage of the power supply. Most failure mechanisms in integrated circuits, for instance, are based on physiochemical reactions that can be accelerated (*stress test*) by temperature in accordance with the Arrhenius equation: $\lambda_b = A \cdot e^{-\frac{E}{KT}}$, where λ_b is the process failure rate, E is the activation energy (in $[eV]$), k is the Boltzmann constant ($8.6 \cdot 10^{-5}$ $[V/K]$), T is the temperature (in $[K]$) and A is a constant. The real value of a given circuit's λ is deduced from $\prod \lambda_{b_i}$) where the b_i identify the components of the circuit.

6.2 Security-oriented Design

Addressing security generally comprises the continuous cycle of risk management and response planning. Because of changes in the environment or the system itself, which may also be time-related, it is impossible to set a fixed strategy. Hence, a cyclic approach to security management [3] has proven most successful. However, a number of guidelines and standards has been developed for security-aware design.

Risk management and the response cycle comprise the following steps: (1) upon detection, a threat agent gives rise to a threat; (2) based on the threat, vulnerability is explored; (3) a discovered vulnerability leads to a risk, which could damage assets or cause exposure, depending on the level/kind of the risk; and (4) the threat must be safeguarded by the threat agent. There are two ways a threat agent can respond to a threat: (1) by mitigation to minimise loss or probability of exposure, or (2) by acceptance with active or passive handling of the security violation. In the latter case, depending on the type of system and the kind of security violation, proper handling is chosen. With intrusions via the Internet active handling means, for instance, to give an intruder bogus results in order to sustain security and discover its identity, whereas passive handling means ending the intrusion process in a controlled way in order to prevent it from causing any (further) damage to the security of the system.

Contingency planning has been developed to systematically integrate security management into technical and business processes. It is composed of four major steps:

1. *(System/Business) Impact Analysis (BIA):* consisting of threat attack identification and prioritisation, business unit analysis, attack success scenario development, potential damage assessment and subordinate plan classification
2. *Incident Response Planning:* comprising incident planning, incident detection, incident reaction and incident recovery
3. *Disaster Recovery Planning:* creating plans for disaster recovery, crisis management and recovery operations
4. *(Business) Continuity Planning:* establishing continuity strategies, plans for continuity of operation and continuity management

The areas of security pertaining to real-time systems can be grouped into the following two main categories:

1. *Environmental:* physical security, personal security, operation security and communication security
2. *Technical:* network security, information security and computer security

Measures coping with environmental threats to security are taken in order to prevent damage to a system either caused by natural disasters and atmospherical effects or the logistics behind the operation or by the personnel. Although very important, these issues are not discussed here — they represent the responsibility of management and security engineers. The issues to be considered here represent the technical aspects of ensuring security of operation.

Technical security issues with computing systems encompass mainly three areas:

1. Computer system security
2. Computer network security
3. Information security

As computer system security the security of a computing system's operation is considered, by taking failure probabilities of, e.g. processors and data storages into account. These issues are handled predominantly by choosing an appropriate architecture for the computing system to allow for its proper operation, even in the case of component failure, by component replication as well as by appropriate failure diagnostics and handling. Here it is of utmost importance not to allow a single point of failure in security-critical designs. Usually the computing system's operation shall be degraded gracefully until a malfunctioning component is repaired or replaced; however, it shall not deny service. Here, real-time restrictions apply to worst-case response times and worst-case repair times.

Computer network security is partly included in computer system security, especially with distributed computing systems. The issues here are mainly the same, except that they involve hardware and software interfaces and drivers, which must ensure data integrity and communication network availability. The basic principle of replication to avoid a single point of failure also pertains to communication networks.

It is only achieved by different means, viz., line replication. Communication lines are usually chosen for transmission speed during normal operation, but for safety during the reduced (on-failure) mode of operation. An issue partly also belonging to information security is here dealt with as well, namely, data security. The information transferred on a network is not only encoded in packages to be correctly processed by the communication drivers, but also encrypted in order to minimise the possibility of extracting information from the data packages by intruders. Real-time restrictions applying here concern network propagation delays and times for re-routing messages on "healthy lines" in the case of broken connection lines.

Information security comprises information integrity as well as confidentiality. Integrity is ensured by the former two areas of technical security, whereas confidentiality is achieved by means of authentication and authorisation. These mechanisms are meant to ensure that only authorised individuals have access to sensitive information like, e.g. records of production processes, medical examinations, or financial transactions. To authorise them, these individuals or organisations need to be authenticated by the data host. Their identity is determined by lock-key mechanisms, which must be safeguarded in order not to reveal the identities to third parties who might misuse the information. The identification mechanisms involve biometrical information (e.g. fingerprints or iris scans) or electronic keys. After an individual is authenticated in a computing system the data flow transferring information across the network is encrypted. The time for authentication needs to be short to prevent brute force attacks. Also the durations of activities need to be monitored, e.g. in order to prevent that someone's data session is taken over by an intruder.

6.2.1 Security Layers

Since security attacks have become more frequent and more sophisticated, investing in single-layer security systems is of limited use. Security experts have built a layered security architecture [2] (see Fig. 6.11) and associated different security management strategies according to their respective strengths and limitations with each individual layer:

Security Policy Layer: this should encompass all aspects of employee awareness of security and responsibility to network, e-mail, Internet and password usage.

Physical Layer: this keeps computers (servers, workstations, portable devices, software media, data back-up media, switches, routers, firewalls, etc.) locked down and safe from physical theft and intrusion. Any device must have a designated owner who is responsible for its security. The owner should have appropriate resources, skills and information to fulfil this responsibility. Network equipment and software should be restricted to authorised personnel.

Data Layer: this is limited to the accessibility of data on any given network. The desired outcome is one that restricts data access to only those users who are required to have access. This protects privacy and provides for accountability. Web application gateways, e-mail spam filters, XML security systems and Secure Sockets Layer (SSL) virtual private networks help to ensure that application traffic is clean, efficient and secure.

Application Layer: this provides an automated security layer to protect configurations on hosts and includes host-based anti-virus applications, intrusion prevention software, spyware tools and personal firewalls. With the most advanced set of tools, these products provide essential "last-resort" security for applications and networks to thwart any potential threat. As with most software, however, human intervention is required to ensure that the solution is constantly updated.

Network Perimeter Layer (NPL): this utilises hardware and conceptual design to provide a layer of protection from outside a network. To create this layer of protection, the NPL utilises firewalls, Virtual Private Networks (VPNs), routers, intrusion detection and prevention software and web content filtering.

Management Layer: this is critical to the continued security of an overall network

- to consolidate the approach to security management, assess the overall vulnerability, and manage patches and updates carefully;
- to persistently monitor all security layers for compliance and vulnerabilities. This includes creating a security framework that makes it possible to identify potential threats early, accurately analyse risks from emerging threats and swiftly develop effective remedial strategies as well as protect an entire organisation, from the borders of the corporate network down to each individual computing system (component). In general, this requires supervision to ensure consistency in assessing overall vulnerability and managing patches and updates for any software and policy.

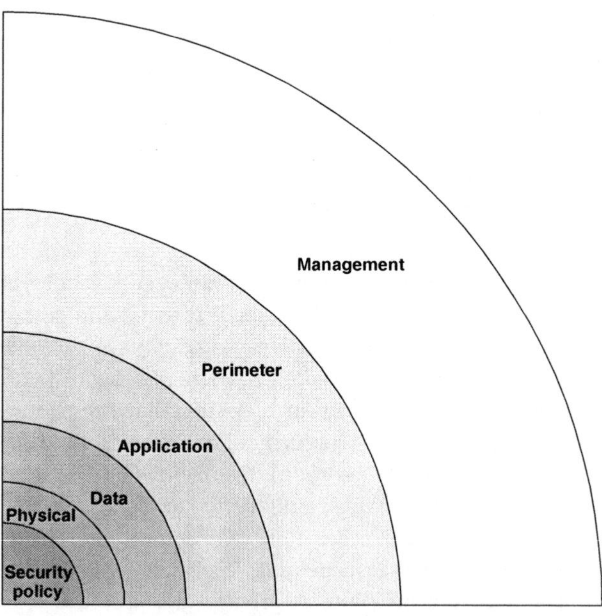

Fig. 6.11. Security layers

From the security point of view, real-time systems do not differ from other computing systems. Sustaining their QoS through security measures is the primary goal, and the tools and methods to achieve this are basically the same. Hence, there is no special notion of real time here, and no distinction is made with respect to the concepts of security. Some security mechanisms do have real-time constraints, but they do not pertain to the systems or the applications. One may establish trade-offs between levels of security and QoS levels, however, since security mechanisms also require computing time and resources, system performance may be affected. In the sequel the most important technical security mechanisms are assessed in more detail.

Encryption: Typical examples of encryption mechanisms for networked systems are Transport Layer Security (TLS) and its predecessor, Secure Sockets Layer (SSL). They represent cryptographic protocols providing secure communications on the Internet (for web browsing, e-mail and other data transfers).

The TLS protocol allows applications to communicate across a network in a way that prevents eavesdropping, tampering and message forgery. It provides endpoint authentication and communication privacy over the Internet using cryptography. Typically, only the server is authenticated (i.e. its identity is ensured), while the client remains unauthenticated. The next level of security – in which both conversation partners are sure with whom they are communicating – is known as *mutual authentication*. It requires Public Key Infrastructure (PKI) deployment to clients unless Pre-Shared Key cypher suits (TLS-PSK) or Secure Remote Password (TLS-SRP) are used, which provide strong mutual authentication without needing to deploy a PKI. The TLS protocol involves three basic phases:

1. Peer negotiation for algorithm support
2. Public key exchange and certificate-based authentication
3. Symmetric cipher encryption

During the first phase, the client and server negotiate cipher suites, which combine one cipher from each of the following:

- Public-key cryptography (e.g. RSA, Diffie-Hellman, DSA)
- Cryptographic hash function (e.g. MD2, MD4, MD5, SHA)
- Symmetric ciphers (e.g. RC2, RC4, IDEA, (Triple) DES, AES, Camellia)

Usually, attacks against a "cryptosystem" without known vulnerability, e.g. Digital Encryption Standard (DES) [64] or Advanced Encryption Standard (AES) [51], are brute force ones trying to find secret keys by exhaustive search in the key spaces. Real-time restrictions apply in order to prevent attackers from being given enough time to "crack" the security keys.

Access Control: Due to the large variety of access control mechanisms from password protection to access control devices (e.g. fingerprint readers, iris scanners, RFID), depending on the level of required security this term does not uniquely identify the access control process. In any case, successful access control encompasses both authentication and authorisation.

The different types of *authentication* mechanisms relate to something you know, something you have, something you are, or something you produce. They have time limitations, as otherwise we would not be able to distinguish authentication requests in the course of a lengthy session, during which an attacker could run a series of access attempts and eventually succeed.

Authorisation must provide services to any authenticated user, members of a group, and/or authorisation across multiple systems. Dial-up protection is used to prevent intrusion by unauthorised individuals, which may use a network's Internet connection to gather information or cause damage. Timing limitations also pertain to authorisation. During a certain period of time, an individual shall receive only one authorisation. Also, during a session, the authorisation system may perform an "inactivity logout" canceling the authorisation after a lengthy period of inactivity. In practice this could otherwise mean an open gate for an intruder accessing the individual's computer during that time.

Since embedded networks also have some Internet connectivity for monitoring or maintenance, these connections are targeted by attackers to collect data from the embedded network or computer systems. In order to detect attackers, scanning and analysis tools may be employed to sense extraordinary executions or information leaks. Intrusion detection systems build on sophisticated authentication and access control algorithms as well as on encryption systems to prevent sensible data leaving intranets, which could be intercepted on the Internet.

Firewalls: In information security, a firewall is any device that prevents a specific type of information from moving between two networks. A firewall may be a separate computer system, a service running on an existing router or server or a separate network containing a number of supporting devices. Access restrictions are applied using authentication mechanisms.

6.3 Concluding Remarks on Design for QoS

All aspects of system design and development discussed are relevant for a designed product's QoS. Although they address different QoS criteria, they all contribute to the overall QoS. A popular fault prevention technique is to introduce *capacity reserves* into the consideration of physical and timing constraints, which reduce the risk of failure. Apart from introducing resource redundancy, the mentioned temporal analysis tools may be helpful to estimate appropriate slack times in the prescribed timing constraints rather than just using rules of thumb based on expert knowledge. In general, however, fault prevention must be viewed in a broader context and dealt with in multiple phases of system development. The concepts of BIT and BIST originate from hardware diagnostics, but their principles can be transferred to software provided that the number of inputs and outputs is limited. In safety-critical environments, the considerations for Safety Integrity Level, life-cycle and fault tolerance mechanisms determine the way systems are devised and maintained. In security-critical applications, the mentioned security measures need to be employed to prevent damage to human health and machinery, or undesired data leakage of sensible information.

7

Design in UML Oriented at QoS

With UML being a prominent design methodology for information systems in general and also being used to design real-time systems, the question arises as to whether UML suggests itself for QoS-oriented design as well. In UML-based projects the following levels of quality are known [74]:

Data quality: accuracy and reliability of data ensuring their integrity
Code quality: correctness of programs and their underlying algorithms
Model quality: correctness and completeness of software models
Architecture quality: quality of a system in terms of its ability to be deployed
Process quality: activities, tasks, rôles and deliverables employed in developing software
Management quality: planning, budgeting and monitoring, as well as the human aspects
Quality environment: all aspects of creating and maintaining the quality of a project, including all above aspects of quality

7.1 UML for Real-time Systems

Two design profiles have been devised for modelling real-time systems in versions 1.4 and 1.5 of UML, namely, RT-UML [17] and UML-RT [66]. These profiles take into account the relevant time-oriented issues in designing real-time systems by introducing necessary concepts in form of new constructs and/or additional diagrams. They explicitly support the consideration of the timeliness issues. One can include the applicable techniques and mechanisms discussed in the previous chapters into designs and the design process.

In the new version UML 2.0, the real-time and QoS-oriented mechanisms, constructs and properties have been introduced in the form of some additional types of diagrams (e.g. timing diagrams, communication diagrams and interaction overview diagrams) and two specialised profiles for timeliness [56] and for QoS and fault tolerance (FT) issues [55].

7.1.1 RT-UML

The profile RT-UML was defined for UML 1.5 to enable the modelling of real-time systems by introducing the following diagrams:

Concurrency diagram (see Fig. 7.1): includes constructs for synchronisation (event flags, e.g. "Router Warm Start", semaphores, e.g. "Position Mutex", signals, e.g. "Emergency Stop"), communication (queues, e.g. "New Container") and shared data (global data or pools, e.g. "New Position") between *active objects*.

Task diagram (see Fig. 7.2): describes relations between *active* and *passive* objects — not classes. Active objects are introduced as compositions of passive objects into process threads.

System architecture diagram (see Fig. 7.3): enhances the *deployment diagram* by representations of a system's inputs/outputs, processing components, and its subsystems joined with its objects and components into a common run-time environment.

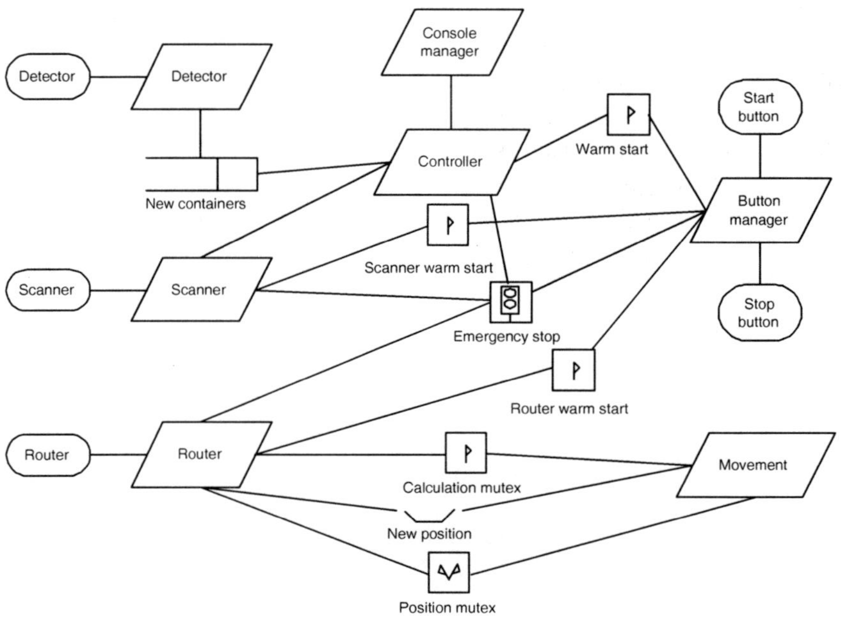

Fig. 7.1. RT-UML concurrency diagram example

The goal of RT-UML was to add as few new basic constructs as possible and, hence, maintain the compatibility of RT-UML with standard UML. As the mentioned three diagram types have been added, and additional non-standard constructs (e.g. mutex or queue) are being used within them, this promise was not entirely kept. One may consider the features, however, as additions to "usual" UML modelling for the special field real-time systems.

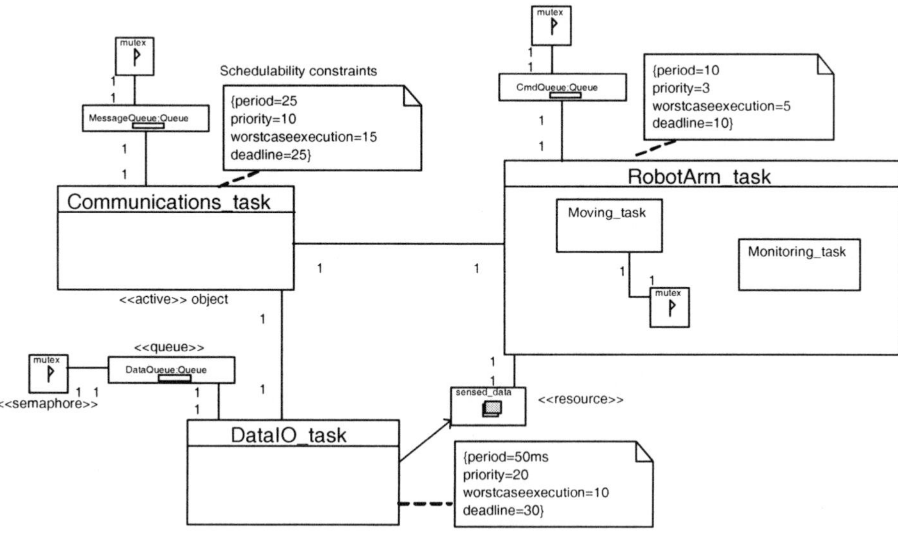

Fig. 7.2. RT-UML task diagram example

7.1.2 UML-RT

The profile UML-RT was devised for UML 1.5 to enable the modelling of real-time systems by introducing a few new basic constructs as stereotypes into UML, making use of its extension mechanisms stereotypes, constraints and tagged values. In particular, the actors known from the ROOM [65] method, being a prominent object-oriented modelling method for real-time systems, were to be made available in UML. The *actor* construct, representing an active object, has been re-named *capsule* in UML-RT, but the semantics remained the same. Being the active objects, capsules are also the only objects appearing in sequence diagrams. In contrast to RT-UML, no new diagrams were introduced with UML-RT.

The *capsule* (see Fig. 7.4) is a specialisation of a *class* with additional semantics and added constraints expressed in the Object Constraint Language (OCL). It represents the active object's class and shares the *class diagram* space with *classes*, representing passive objects. Its attributes and operations have the same place and rôle. Added to the description are *ports* through which capsules interact with one another and with the environment. Capsules can be nested, i.e. a capsule may contain one or more other capsule(s) (rôles). The structure and state diagrams of capsules represent their structures and interconnections as well as life-cycles, respectively. Each interconnection has its associated protocol, and two interconnected ports have to share the same *protocol* in order to be able to "talk" with each other. The signals that are transferred between them drive the capsules' inherent *state charts* that represent their life-cycles. Each capsule can become the "top-capsule" to a component, meaning that it initiates and drives the execution at the processing node.

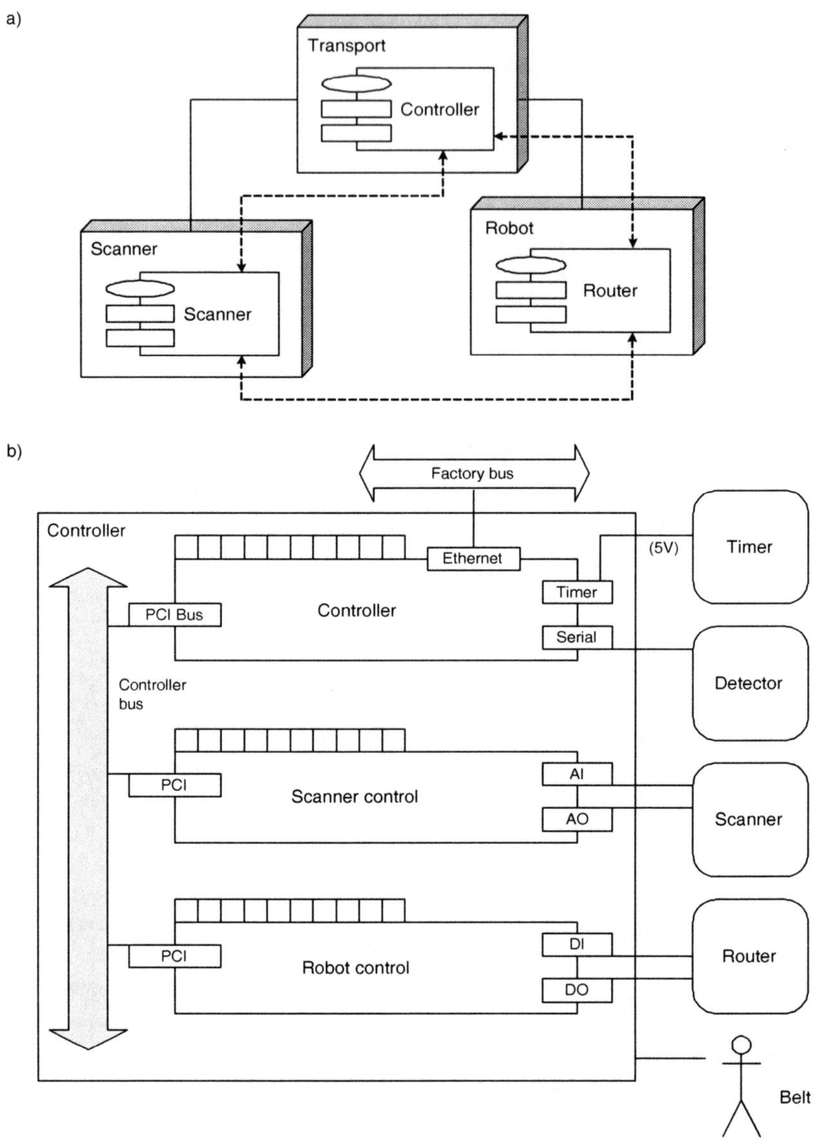

Fig. 7.3. An RT-UML system architecture diagram (b) for the example enhances its software deployment diagram (a)

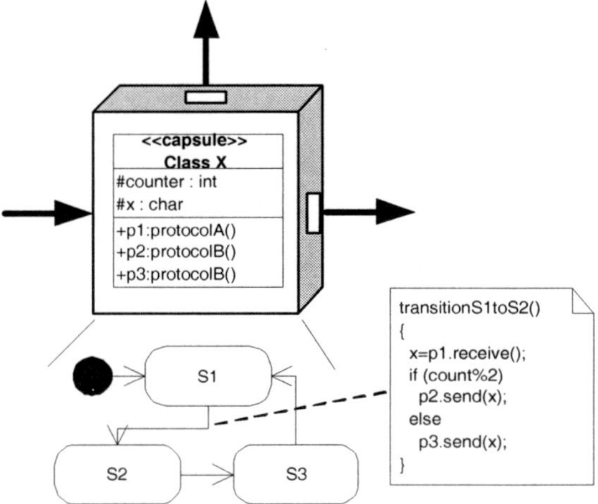

Fig. 7.4. UML-RT capsule semantics

7.1.3 UML 2.0

In the previous two sections we saw two different approaches to enhance UML with constructs that enable the expression of real-time features. In RT-UML, new diagrams have been introduced, while keeping the original basic constructs, and in UML-RT, new basic constructs have been added as stereotypes to the existing ones. In UML 2.0 none of the two was sustained as such. In Fig. 7.5 the new types of diagrams introduced in UML 2.0 are highlighted.

For architecture modelling in UML 2.0 capsule-like "Structured Classes" have been introduced [57] also supporting the Model Driven Architecture (MDA) approach. Like a capsule, a structured class has its associated "(Composite) Structure Diagram" (e.g. Fig. 7.6) showing its structure — "parts" ("capsule rôles" in UML-RT) representing objects used by an instance of the class and possibly created and destroyed with that instance. It also shows the "connectors" representing the relationships between these parts. Structured classes may posses "ports" (like "capsule ports" in UML-RT) that facilitate communication between them. All interactions (calls or asynchronous messages) go through them. They are optional, and calls may also be made straight to objects. A port can have two types of interfaces: a "provided interface" is a set of operations the port makes available to the outside world, while a "required interface" is a set of operations the port may use on the outside world. It can have zero or more of each type of interfaces. Port compatibility is statically verified: each port at one end of a connector must provide every operation that is required by the port on its other end, i.e. expressed in UML-RT terminology, the connected ports have to share the same "protocol". In UML 2.0 components resemble structured classes very much (see Fig. 7.7), so we may also find ports and interfaces (implemented as classes) here. The decision which decomposition level is the most suitable depends on the problem being modelled.

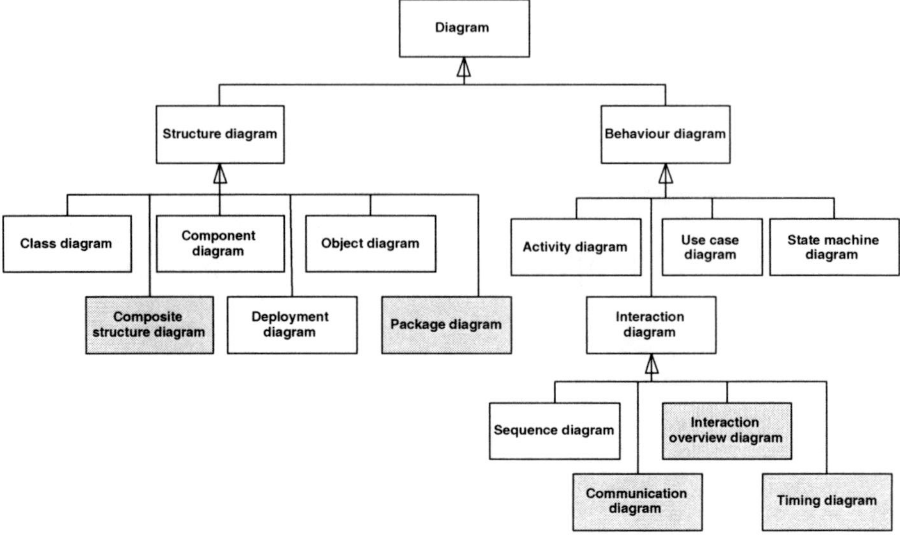

Fig. 7.5. UML 2.0 diagram structure

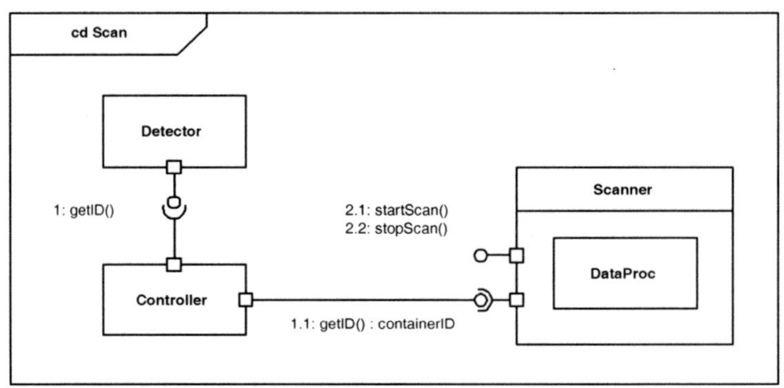

Fig. 7.6. Fragment of a composite structure diagram

To support the expression of real-time features, *timing diagrams* have been introduced, enabling the representation of temporal interdependencies between active objects as well as the expression of their synchronisation and communication interdependencies. They are reminiscent of Gantt or "bus timing diagrams". The "state lifeline" is used to display the change of state over time, where the X-axis displays elapsed time in chosen time units, and a given list of states is noted on the Y-axis (see Fig. 7.8). The "value lifeline" shows the change of values over time, where the X-axis displays elapsed time again, while value changes are depicted by over-crossings of two horizontal lines (see Fig. 7.9). Temporal constraints, duration constraints as well as events triggering state/value changes can be drawn on the "lifelines". State and value lifelines can be stacked one on top of the other in any combination as far

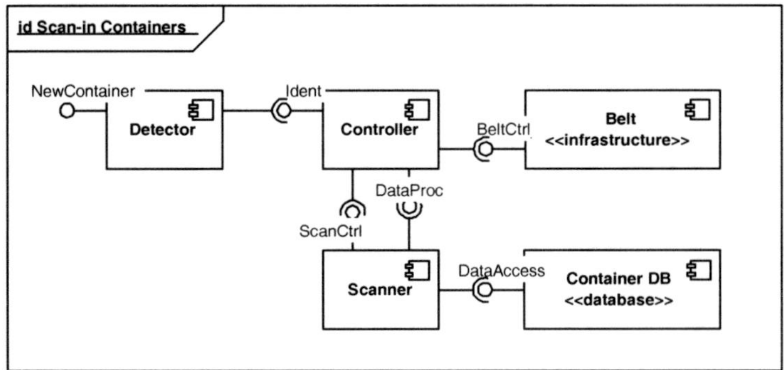

Fig. 7.7. An example component diagram

as they share the same X-axis (e.g. Fig. 7.10). Messages (events) can be passed from one lifeline to another, which is shown in form of arrows.

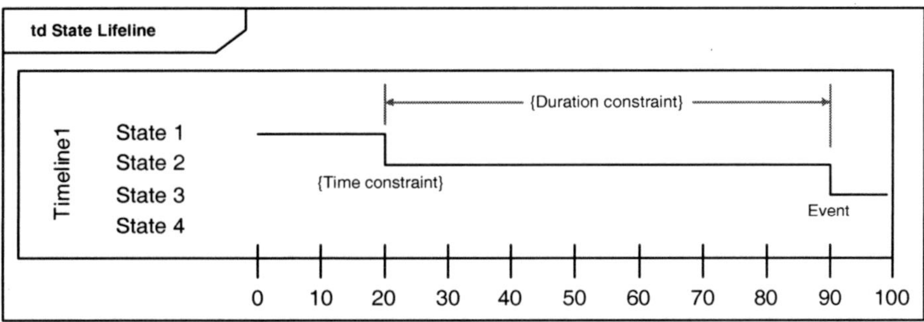

Fig. 7.8. Timing diagram structure: state lifeline

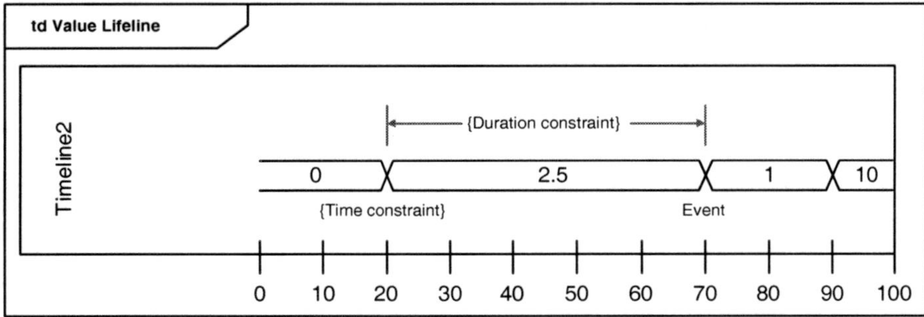

Fig. 7.9. Timing diagram structure: value lifeline

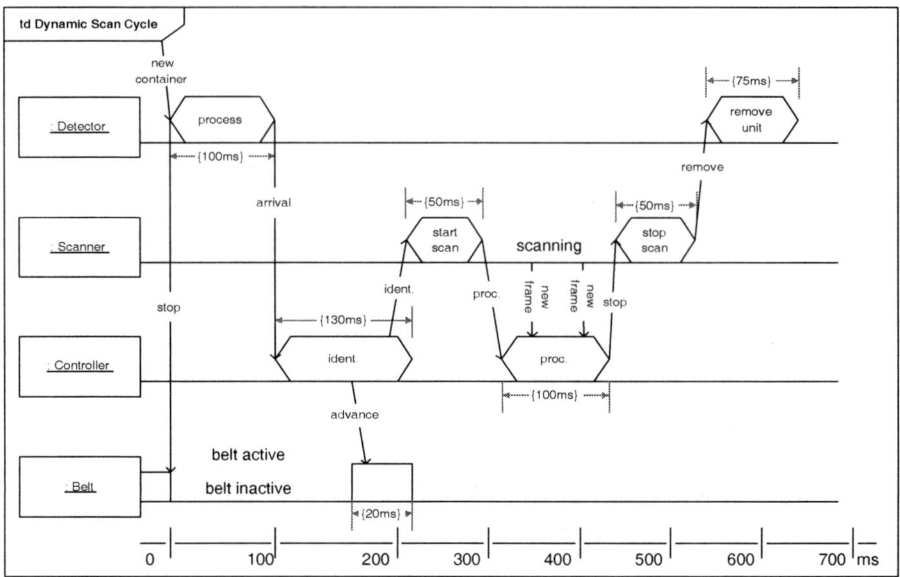

Fig. 7.10. An example timing diagram: lifelines joined on a common time-scale

For the hierarchical representation of control flow on the project level of design *interaction overview diagrams* can be used instead of *activity diagrams*. An interaction diagram consists of interaction diagram nodes, which can include sequence, communication, interaction overview and timing diagrams. The common connecting constructs of activity diagrams are used in interaction overview diagrams in the usual way (e.g. initial, final, decision, merge, fork and join). There are two kinds of interaction diagram nodes: interaction occurrences representing references to interaction diagrams and denoted by "ref" in the top left corner of the corresponding frame, and interaction elements which display the contents of an interaction diagram node in line with its name in the top left corner of the frame (see Fig. 7.11).

To represent relations and data flows not only between objects but also their collections *communication diagrams* were introduced in place of *collaboration diagrams*. They resemble sequence diagrams. In communication diagrams (see Fig. 7.12), objects are shown with association connectors between them. Messages are added to the associations and shown as short arrows pointing in the direction of message flow.

For a more detailed representation of real-time and QoS issues, two profiles with appropriate stereotypes and patterns have been defined and can be used in UML 2.0 projects. They are the subject of the two following sections.

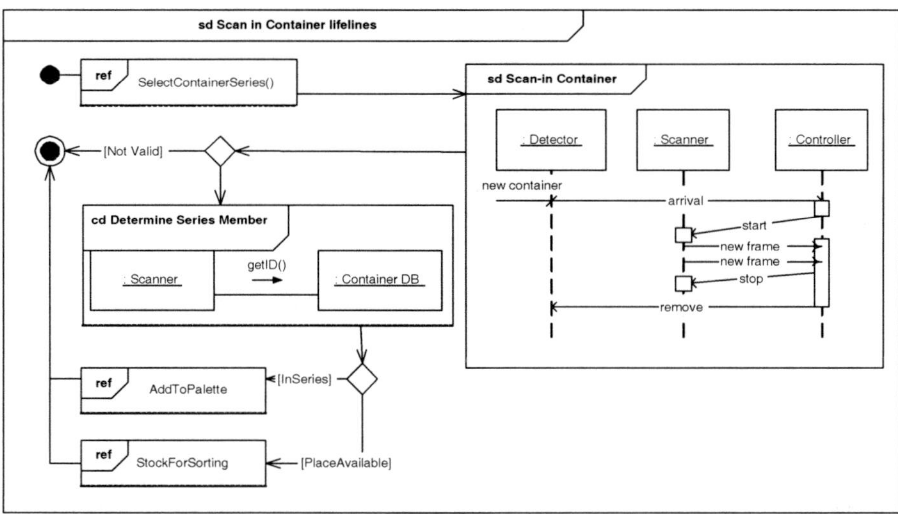

Fig. 7.11. An example interaction overview diagram

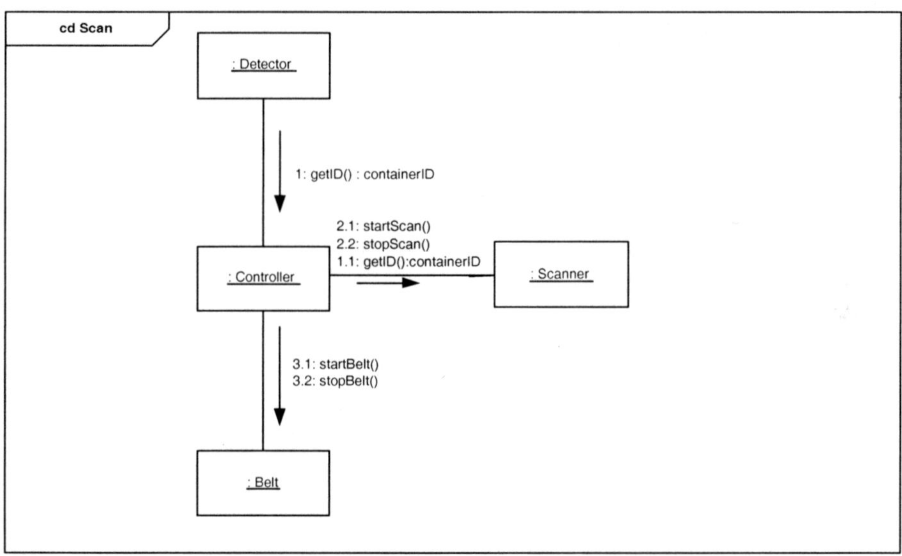

Fig. 7.12. An example communication diagram

7.2 UML Profile for Schedulability, Performance and Time Specification

The overall approach taken in the profile for schedulability, performance and time specification [56] was not to predefine a canonical library of real-time modelling concepts. Instead, modelers are left the freedom to represent their real-time solutions in a way most suitable to their needs. In addition, the following guidelines have been followed:

- The profile does not change UML unless absolutely necessary
- It does not restrict the use of UML in any way
- It introduces stereotypes (e.g. resources) and tagged values (e.g. timing and QoS properties), which should enable quantitative analysis, into UML diagrams
- It supports synthesis and analysis of models without requiring knowledge on analysis techniques such as queuing theory or Rate Monotonic Analysis

The profile for schedulability, performance and time is partitioned into a number of subprofiles, i.e. profile packages dedicated to specific aspects and model analysis techniques (see Fig. 7.13). At the profile's core is the set of subprofiles that represent the general resource modelling framework, introducing resources and their QoS properties. These provide a common base for all analysis subprofiles in this specification. It is anticipated that (future) profiles, dealing with other types of QoS (e.g. availability, fault tolerance, security), may need to re-use only a portion of this core. Hence, the general resource model itself is partitioned into three separate parts:

1. *RTresourceModeling:* resource model
2. *RTtimeModeling:* time modelling with classes *Clock, Timer, TimeValue*
3. *RTconcurrencyModeling:* model of parallelism with scenarios and stimuli

The parameters for scheduling scenarios and resources are collected in the framework of the Analysis Models package, containing subprofiles for schedulability analysis (*SAProfile*) and performance analysis (*PAProfile*). The modular structure shown in Fig. 7.13 allows to use only the subset of the profile really needed, i.e. a particular profile package and the transitive closure of any profiles that it imports (e.g. a user interested in performance analysis would need the packages *PAProfile, RTtimeModeling* and *RTresourceModeling*.

In addition, in this specification two technology-specific subprofiles have been included: a model library containing a high-level UML model of Real-Time CORBA (*RealTimeCORBAModel*) and its associated schedulability analysis profile (*RSAProfile*). They should enable the construction of complex models, for which it is important not only to model the applications but also the supporting infrastructure – in this case, the infrastructure provided by Real-Time CORBA [54].

7.2.1 General Resource Modelling Framework

The essential framework for modelling real-time systems using UML is the General Resource Modelling framework (GRM). Its central concept is the notion of a

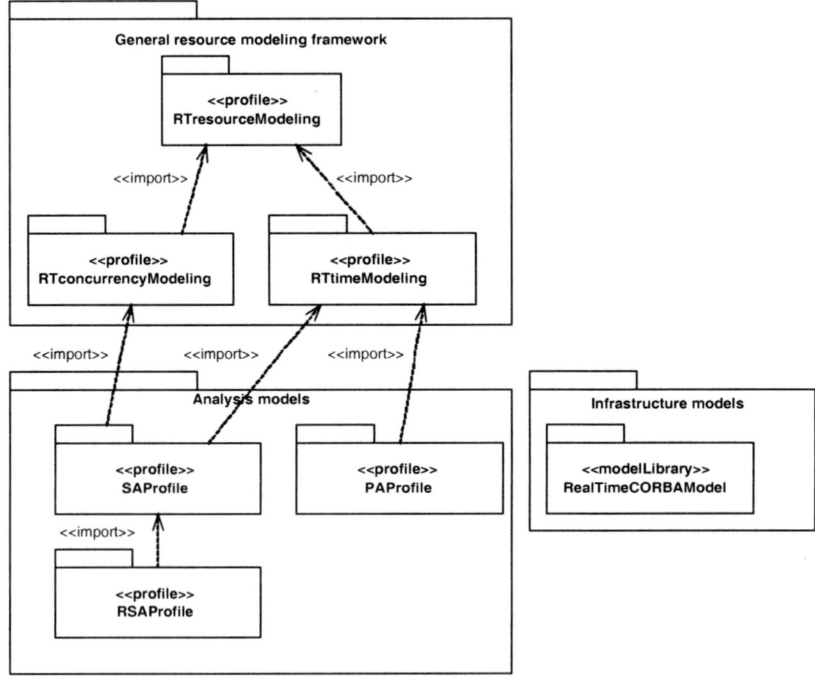

Fig. 7.13. Structure of the profile

resource instance. The latter represents a run-time entity that offers one or more services, for which a measure of effectiveness or quality of service needs to be expressed. Such a measure might, for instance, be the execution time of the service or its precision. The fact that it is necessary to specify the quality of service is simply a reflection of the finite nature (limited capacity, limited speed, etc.) of the resources physical underpinnings. Figure 7.14 deals with the different categorisations of resources that are useful to model real-time systems. A given resource instance may belong to more than one type, although it can be classified by at most one type from each category.

7.2.2 General Time Modelling

The time domain model identifies the set of time-related concepts and semantics that are supported, directly or indirectly, by this profile. It is partitioned into the following separate, but related groups of concepts:

- Concepts to model time and time values
- Concepts to model events in time and time-related stimuli
- Concepts to model timing mechanisms (clocks, timers)
- Concepts to model timing services, such as the ones found in real-time operating systems

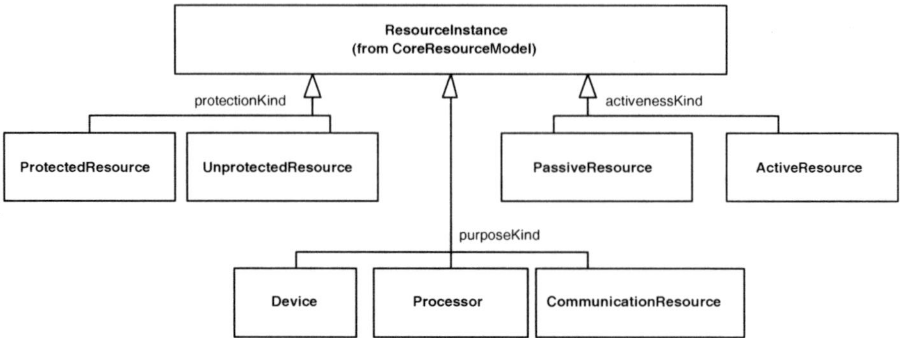

Fig. 7.14. Model of resource types

There are two basic types of timing mechanisms in the domain model:

1. *Timers* are mechanisms that may generate time-out events when specified time instants occur; these may either be the instants when some clocks reach predefined values, or when prespecified time intervals have expired relative to given instants (usually the instants when timers are started).
2. *Clocks* are mechanisms periodically causing clock tick events to occur; clock ticks may, in turn, cause stimuli called clock interrupts.

There are two ways to specify time values using this profile, viz., by using the $<<$ *RTtime* $>>$ stereotype to identify model elements that represent time values, and by using instances of the Tagged Value Language's (TVL) data type *RTtimeValue* (or its subclasses), which is defined in this profile. A time interval ($<<$ *RTinterval* $>>$ stereotype) can be represented either as an absolute interval, in which case it has both a start and end time specified by *RTintStart* and *RTintEnd* tags, respectively, or as a duration (*RTintDuration* tag), in which case it is a relative time interval. The values of these tags are instances of the TVL data type *RTtimeValue*. The stereotype $<<$ *RTtimingMechanism* $>>$ is defined as an abstract one capturing the common QoS characteristics of timers and clocks (e.g. stability, drift, skew, maximum value, origin, resolution, offset, accuracy). It is not intended to be employed by modelers, who are expected to use either $<<$ *RTtimer* $>>$ or $<<$ *RTclock* $>>$.

A time-bounded activity can be modelled by applying the $<<$ *RTaction* $>>$ stereotype to any model element that represents an action execution or its specification (e.g. methods, actions including state actions, entry and exit actions of state machines, states, subactivity states and transitions). The start and end times of the action are specified by appropriate tagged values (*RTstart* and *RTend*, respectively). Alternatively, they may be assigned the tag *RTduration*.

Since event occurrences do not show up explicitly in UML models, the stereotype $<<$ *RTevent* $>>$ can be applied to any model element that implies an event occurrence, such as action executions, methods, state actions including entry and exit actions of state machines, states, subactivity states, and transitions. It can also be applied to stimuli (and their descriptors) to model stimuli taking time to arrive at their destinations. In all these cases, it specifies the start time of the associated behaviour. Thus, it can be used as an alternative to timed actions where the durations or end of the actions are not significant. Alternatively, to model any stimulus that has an associated time-stamp (*RTstart* and *RTend*), the stereotype $<< RTstimulus >>$ can be used. Specialised examples of the latter are the stereotypes $<< RTclkInterrupt >>$, $<< RTtimeout >>$, and $<< RTdelay >>$. An example of using these stereotypes is shown in Fig. 7.15.

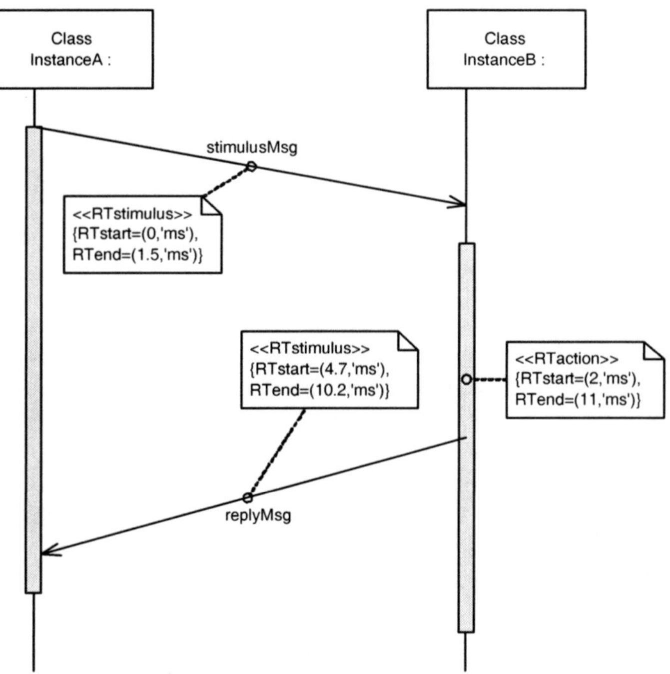

Fig. 7.15. Time annotations in a sequence diagram

7.2.3 General Concurrency Modelling

The domain model of the general concurrency model is shown in Fig. 7.16. A concurrently executing entity is indicated by stereotyping any resource (classifier or instance) with the stereotype $<< CRconcurrent >>$. The main thread of the active instance is represented by a tagged value on a classifier referencing a method; it is always derived for a concurrent instance, its classifier.

Causality within procedures is largely dealt with using standard UML concepts, such as Action Sequence and Action. However, methods need to be linked to actions, which is represented by a stereotype ($<< CRcontains >>$) of the standard usage relationship. Causality between procedures is dealt with by stereotyping both actions

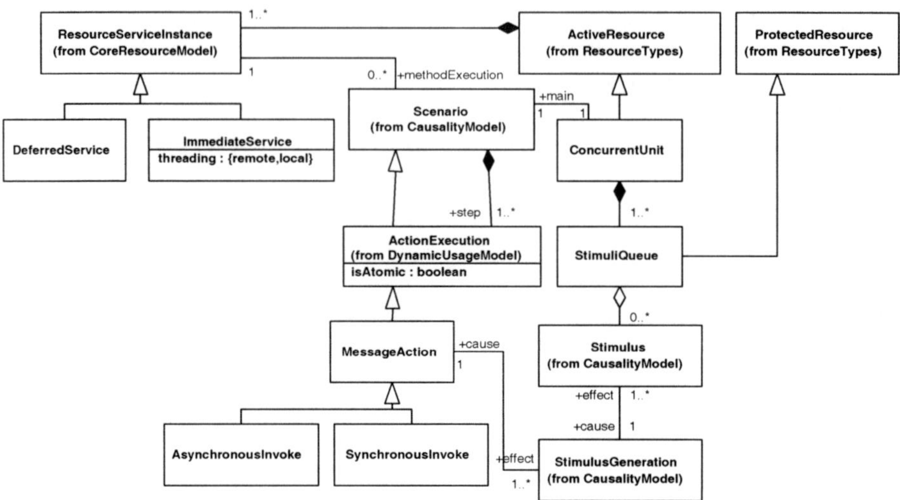

Fig. 7.16. Concepts of general concurrency modelling

(in this case Call and Send), and Operations and Receptions (or Messages and Stimuli as convenient). Calls and Sends may be stereotyped as either $<< CRasynch >>$ or $<< CRsynch >>$ to indicate whether they wait for a response or not. Operations and Receptions may be stereotyped as either $<< CRimmediate >>$ or $<< CRdeferred >>$ to indicate how messages are dealt with. In the case of immediate handling, the indication of threading is dealt with by a tagged value. The linkage between message action and handling procedure is achieved using standard UML associations. Atomicity of actions is dealt with by applying the stereotype $<< CRaction >>$ and setting the (Boolean) tagged value *CRatomic*.

Figure 7.17 shows a sequence diagram where the various model elements have been stereotyped from the concurrency model. It shows how data are gathered and displayed in the data acquisition system, which also features some asynchronous communication.

7.2.4 Schedulability Modelling

Scheduled jobs are the elementary parts of applications competing for the use of resources that execute schedules, i.e. processors more generally called execution engines. By a *schedule* an assignment of all jobs in a system to be carried out on available execution engines is meant, which is produced by the scheduler. For systems of execution engines, a schedule is the union of the schedules for the individual execution engines. Jobs are scheduled and any resources required are allocated according to a chosen set of scheduling algorithms and resource arbitration policies. The component that implements these algorithms is called *scheduler*. A *scheduling mechanism* defines an implementation technique used by a scheduler to make deci-

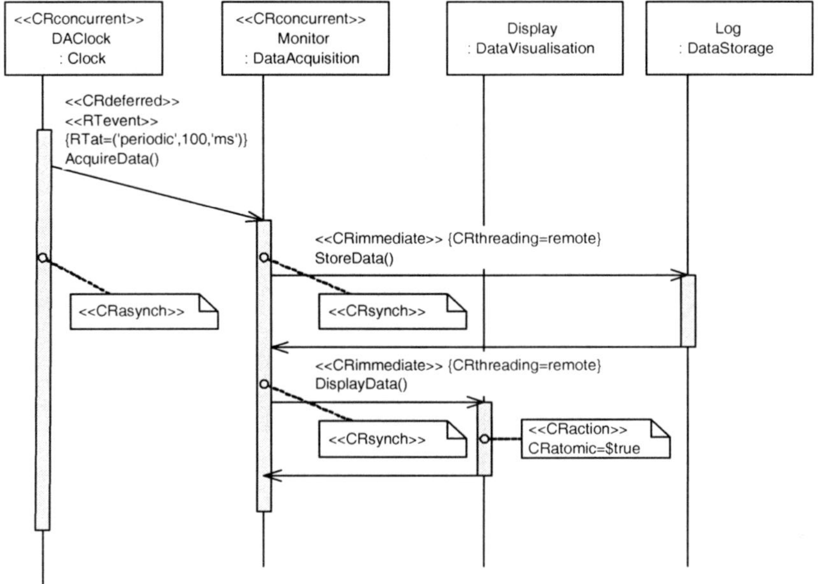

Fig. 7.17. Concurrency model of the asynchronous case

sions about the order of execution. To analyse the schedulability of a system, three schemes must be defined:

1. A scheduling policy or algorithm by which the order of processing is determined
2. A resource arbitration scheme (allocation control policy) by which allocation decisions for resources are made
3. A model analysis method by which parameters that affect scheduling and resource usage can be assigned

Figure 7.18 depicts a general schedulability model that identifies the basic abstractions and relationships used in schedulability analysis. A *real-time situation* is modelled as a stereotype << *SAsituation* >> of a UML collaboration (or a set of collaboration instances). This means that all interactions specified in that collaboration represent scheduled jobs. A scheduled job maps to an interaction (or a set of interaction instances). It is represented by a trigger, which is used to stereotype the first message (stimulus) in the interaction, and a response, which is used to stereotype the associated action (action execution) of the trigger. The schedulable resource used by the response to execute can be derived from the action's "owner" with the stereotype << *SAresponse* >>. Resources and execution engines are indicated by stereotyping classifier rôles (instances) with the appropriate stereotypes (<< *SAengine* >>, << *SAschedRes* >>, << *SAresource* >>). The relationship between execution engines and "shareable" resources is established using the stereotypes << *SAowns* >> and << *GRMdeploys* >> of a realisation (i.e. abstraction).

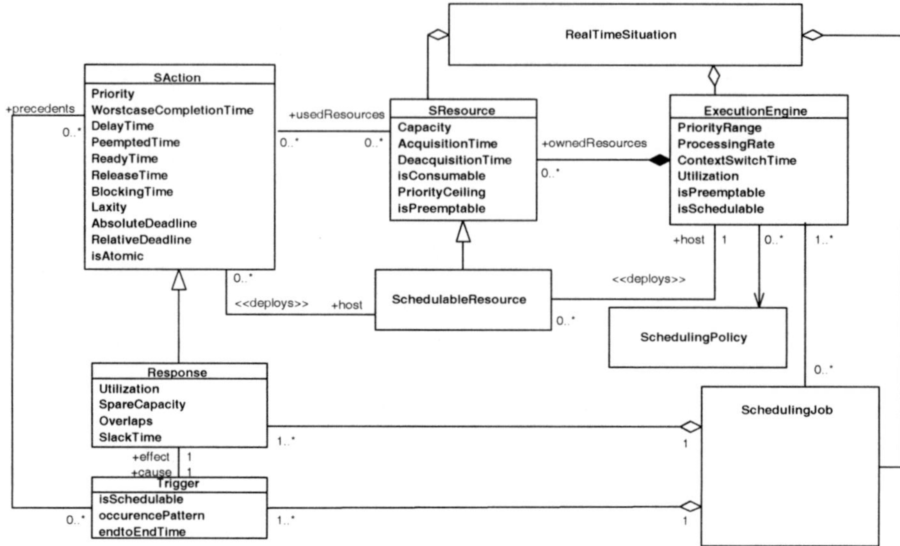

Fig. 7.18. The core schedulability model

Actions are either stereotyped $<< SAaction >>$s (action executions) or, for diagrammatic convenience where appropriate, stereotyped messages (stimuli). The "uses" relationship can often be derived from the destination of a message, but in cases with coarse levels of granularity and where messaging actions are not being used explicit usage dependencies (stereotyped $<< SAuses >>$) may be inserted between the appropriate actions (action executions) and resources. A precedence relationship among jobs may potentially be modelled using an *Interaction* between scheduled jobs, in which case any action preceding a stimulus message tagged as an $<< SAtrigger >>$ would indicate the presence of a "precedes" association. It was, however, decided to represent any scheduled job as a separate *Interaction* and, hence, an explicit usage dependency among jobs ($<< SAprecedes >>$) is necessary. In Fig. 7.19 a sequence diagram of the example data acquisition system is shown presenting detailed descriptions of triggers and responses and introducing the discussed concepts.

7.2.5 Performance Modelling

The part of the profile intended for general performance analysis of UML models provides facilities to:

1. Capture performance requirements within design contexts
2. Associate performance-related QoS characteristics with selected elements of UML models
3. Specify execution parameters that can be used by modelling tools to compute predicted performance characteristics

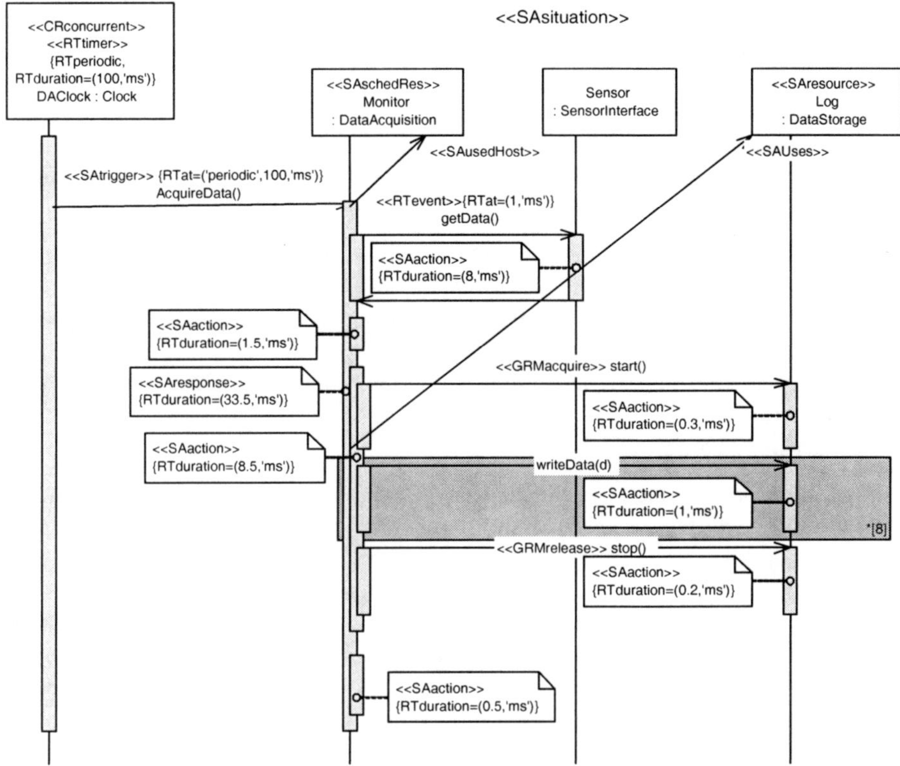

Fig. 7.19. Triggers and responses in a sequence diagram

4. Present performance results computed by modelling tools or found in testing

Scenarios define response paths whose end points are externally visible, so they represent responses with response times and throughputs. On scenarios QoS requirements are placed. In performance-related models, any scenario is executed by a *job class* or *user class* with an applied load intensity, and these classes are either open or closed. To avoid confusion with software classes, here such a class is called a *workload*. An open workload has a stream of requests that arrive at a given rate in some predetermined pattern (such as Poisson arrivals), while a closed workload has a fixed number of active or potential users or jobs that change between executing the scenario and spending a *delay period*, sometimes called *think time* and determined outside of the system between the end of one response and the next request. Scenario steps or activities are the elements of scenarios, and are joined in sequences. Any scenario step may optionally have its own QoS properties.

Resource demands by a scenario step include its host execution demand as already mentioned and the demands of all its substeps. Resources are modelled as servers. The *service time* of an active resource is defined as the host execution demand of the steps it hosts. Thus, different service classes, given by different steps, may have different service times. This places the definition of service times squarely within software specifications. A device may have, however, a speed factor that scales all the steps running on that resource.

Performance measures for a system include resource utilisations, waiting times, execution demands (for CPU cycles or seconds) and response times (the actual time to execute a scenario step or scenario). Each measure may be defined in different versions, several of which may be specified in the same model, such as:

1. A required value, coming from the system requirements or from a performance budget based on them (e.g. a required response time for a scenario)
2. An assumed value, based on experience (e.g. for an execution demand or an external delay)
3. An estimated value, calculated by a performance tool and reported back into the UML model
4. A measured value

Further, a reported value is one of several possible statistical properties of a given measure, such as its average, maximum or x-th percentile (e.g. the 90th percentile meaning that 90% of values are smaller than this).

Since performance is a dynamic property, scenarios play a key rôle in determining a systems performance characteristics from its UML models. In UML, scenarios are most directly modelled either using collaborations or activity graphs. A performance context is modelled as a stereotype $<< PAcontext >>$ of a UML collaboration (or a set of collaboration instances). In collaboration-based scenarios, a scenario maps to an interaction (or a set of interaction instances) and is represented by its first (root) step. A step represents an execution of some action. There are two alternatives to identify performance steps in collaborations, viz., associating a step stereotype ($<< PAstep >>$) directly with an action execution model element, or associating the stereotype with the message (stimulus) model element that directly causes that action execution. If the step is a root step, then it may optionally be stereotyped with the appropriate workload stereotype $<< PAopenLoad >>$ or $<< PAclosedLoad >>$. In activity graph-based scenarios, a performance context can be modelled by an activity graph that is stereotyped as a $<< PAcontext >>$. Scenarios are modelled by the set of states or activities and transitions of the activity graph. Each action or subactivity state of the graph is stereotyped as a $<< PAstep >>$.

Processing resources can be modelled in two different ways. The direct one is to associate the appropriate stereotype $<< PAhost >>$ with any classifier rôle (or instance) executing scenario steps. This is only useful, however, in cases where each classifier rôle (instance) is executing on its own host. More common is the situation that different classifier rôles (instances) are executing on different hosts with a great probability that some hosts will be hosting more than one rôle (instance). Then, the collaboration does not contain sufficient information to determine the allocation of rôles (instances) to hosts. Hence, it is necessary to determine which processor resource is running which classifier rôle (instance) on the basis of $<< deploys >>$ relationships to nodes that are stereotyped as hosts. Passive resources are represented directly by classifier rôles (or instances) stereotyped with the $<< PAresource >>$

stereotype. Alternatively, they may be modelled indirectly by specifying one or more *PAextOp* tagged values for the steps that access these resources.

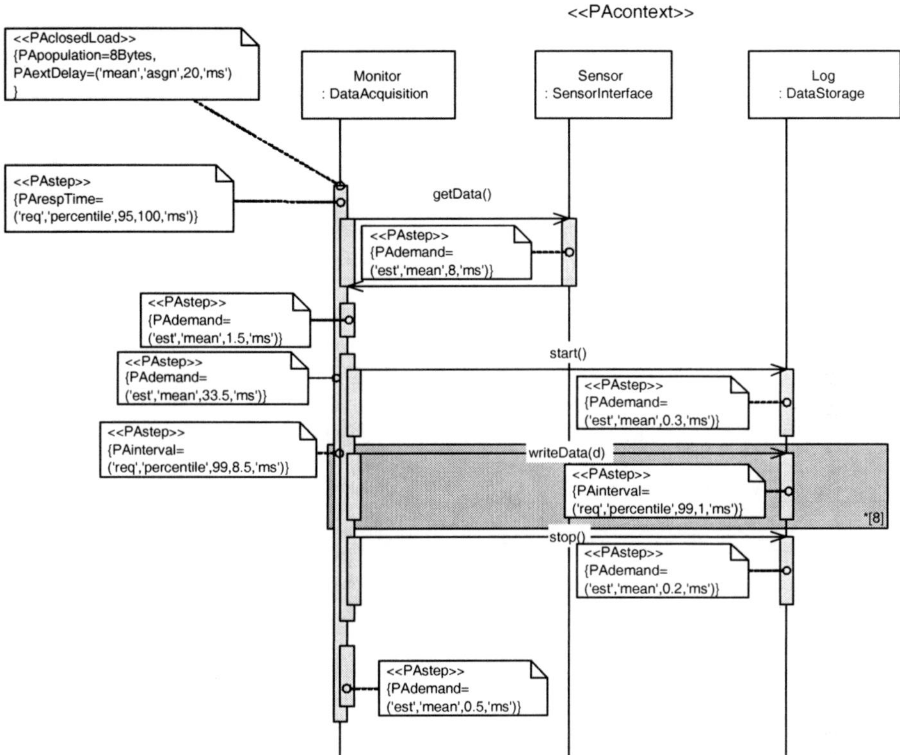

Fig. 7.20. Sequence diagram of the data acquisition example with performance annotations

In Fig. 7.20 a sequence diagram for the data acquisition scenario is outlined. The QoS attributes associated with the events and actions in this diagram are taken directly from the requirements. The closed workload is associated with the first action execution, since there is no explicit user in the model (had there been one, the workload could have been attached to the stimulus coming from the user to the monitor).

7.2.6 Applications of Real-Time CORBA

Real-Time (RT) CORBA is a set of extensions built on top of the standard Common Object Request Broker Architecture (CORBA). One of its primary purposes is to provide a framework for predictable end-to-end execution times of CORBA-based applications in environments that may be distributed and heterogeneous (e.g. combining several different parts based on real-time operating systems into a common application).

The basic CORBA model, also supported by RT-CORBA, is client-server. A client makes a series of requests to one or more servers that may be in address spaces different of the client's. In the real-time context, such execution scenarios ("activities" in RT-CORBA) typically have deadline constraints. Since there can be multiple scenarios running concurrently, possibly sharing some of the same servers and resources, conflicts can arise so that special actions need to be taken to ensure that deadlines are respected. This requires the knowledge of relative priorities of activities, controlled access to shared resources (to prevent deadlocks and to minimise priority inversion), etc. RT-CORBA provides built-in mechanisms to handle this, provided that it is given information on the QoS characteristics of the various scenarios.

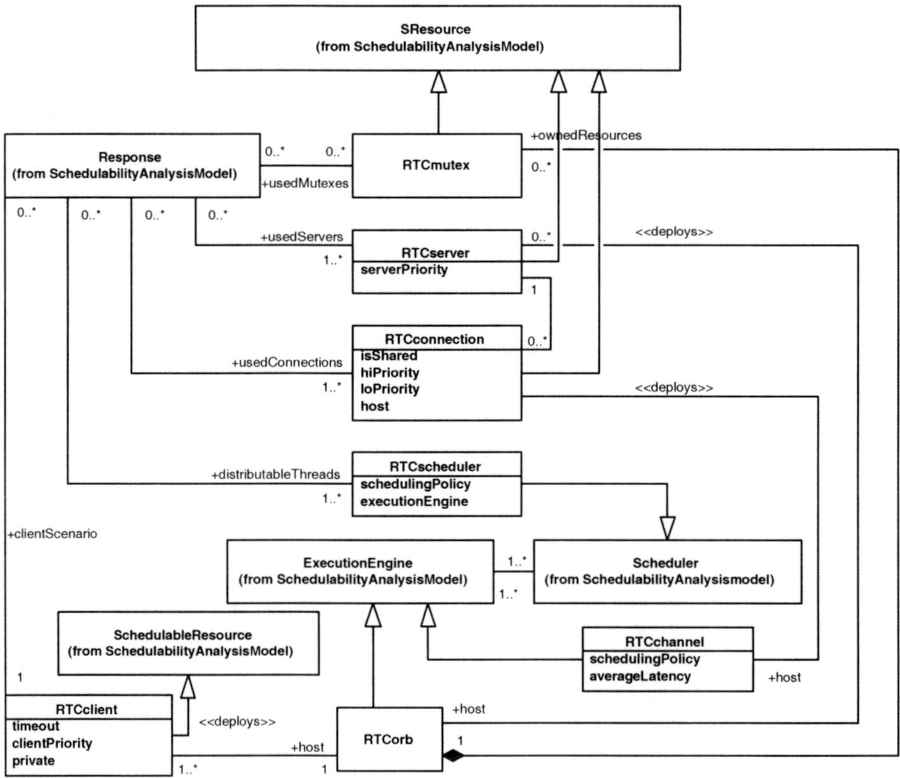

Fig. 7.21. Domain model of Real-Time CORBA

Figure 7.21 depicts a domain model of RT-CORBA applications. The model is fully based on the conceptual framework provided by the scheduling domain model (see Fig. 7.18). This approach was chosen to simplify the schedulability analysis of UML models that represent RT-CORBA applications. Since it is a specialisation of the schedulability subprofile, the approach used is identical. All extension elements defined for this subprofile carry the prefix "RSA", which stands for Real-Time CORBA Schedulability Analysis.

7.2.7 The MARTE Profile

The MARTE profile [58] builds on the mentioned Profile for Schedulability, Performance and Time [56], and is aimed at Modelling and Analysis of Real-Time Embedded systems (MARTE). It provides support for specification/design and verification/validation, and is intended to eventually replace the mentioned profile.

The MARTE profile supports more detailed design of hardware and software structures as well as annotating software constructs with information meant for schedulability analysis. This may enable interoperability between tools used for design, verification, code generation, etc., on one hand, and for decision making from the integral properties of hardware and software components based on model analyses on the other.

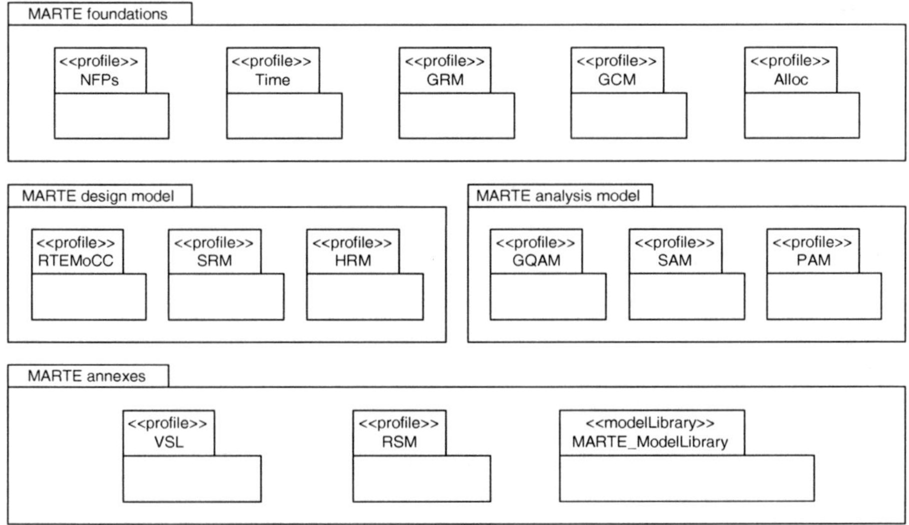

Fig. 7.22. Structure of the MARTE profile

The structure of the profile is shown in Fig. 7.22, where the acronyms stand for Non-Functional Properties (NFPs), Generic Resource Modelling (GRM), Generic Component Model (GCM), Allocation modelling (Alloc), RTE Model of Computation and Communication (RTEMoCC), Software Resource Modelling (SRM), Hardware Resource Modelling (HRM), Generic Quantitative Analysis Modelling (GQAM), Schedulability Analysis Modelling (SAM), Performance Analysis Modelling (PAM), Value Specification Language (VSL) and Repetitive Structure Modelling (RSM), respectively.

7.3 UML Profile to Model QoS and FT Characteristics and Mechanisms

The UML profile for QoS and fault tolerance [55] enables consistency in describing various QoS properties in UML models. It also addresses some features of the profile for schedulability, performance and time by specialising some QoS-oriented classes from the GRM.

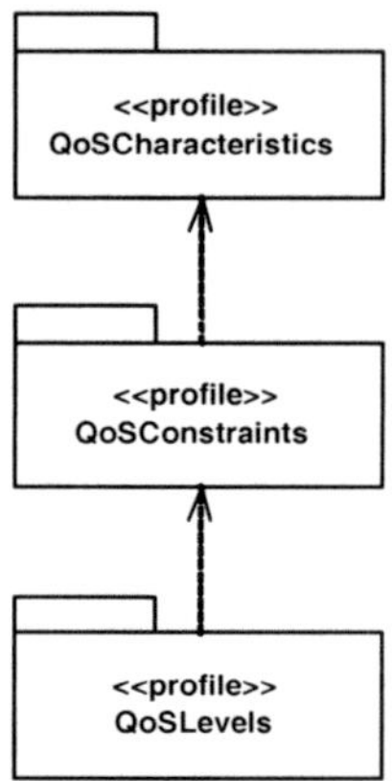

Fig. 7.23. Metamodels of the QoS profile

The structure of the QoS profile corresponds to the metamodel structure shown in Fig. 7.23, which introduces the elements required to model QoS systems. The package *QoS Characteristics* includes modelling elements to describe QoS characteristics, their *Categories* and *Dimensions*. The package *QoS Constraints* comprises modelling elements to describe constraints (limits of QoS characteristics) and QoS contracts (agreements between provider-*QoSOffered* constraints and client-*QoSRequired* constraints; see Fig. 7.24). The package *QoS Levels* contains modelling elements to specify QoS modes and transitions. Based on the working scenarios and configurations of the (sub) systems, different working modes can be assigned to the components, in which they provide the same services in different qualities. For each working mode of a (sub) system, a QoS level can be specified for its *Allowed Space* of QoS constraints. When it leaves this scope, a transition is triggered to another QoS level. The QoS levels define the QoS behaviour of model elements. Of this behaviour it is expected that, when the quality state is at some quality level, the QoS constraints associated with the level must be fulfilled. The profile defines hierarchically decomposable QoS constraints from which the QoS requirements and characteristics of a system are derived, meeting a contract between them (see Fig. 7.24).

The general categories of QoS characteristics that can be expressed in the profile are:

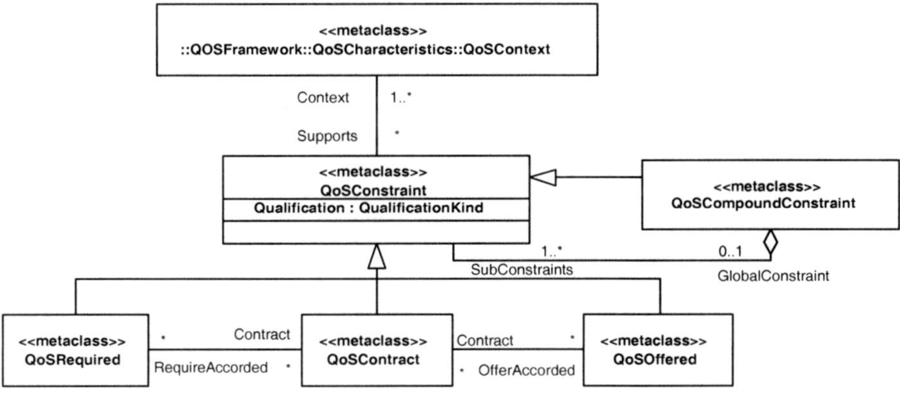

Fig. 7.24. QoS constraints

Performance: this addresses timeliness in terms of latency (interval in which event responses must arrive), throughput (number of event responses handled) and utilisation, where schedulability and performance profile is addressed

Dependability: reliance on a service, i.e. availability (proportion of "up-state" time), reliability (maintenance of the specified level of performance), safety and integrity (e.g. "graceful degradation" also contributes to ensuring integrity from the dependability point of view)

Security: access control and confidentiality

Integrity: is the service provided also the service expected?

Coherence: concurrent and temporal consistency (expressed in terms of throughput and latency)

Demand: how much of a resource is needed (addressing the profile for schedulability, performance and time)?

The listed QoS properties are assembled in packages forming the QoS profile based on their categories (see Fig. 7.25).

The fault tolerance (FT) extensions of UML comprise the *FT architectures metamodel* (see Fig. 7.26) and support the following features:

1. No single point of failure
2. Systematic monitoring and control of applications
3. Capability for manual system control (e.g. start/stop or degraded mode)
4. Global and redundant supervision
5. Capability for local supervision
6. Widely used synchronisation algorithms and tolerance to loss of external time reference

The packages of the FT profile include stereotypes to describe four main concepts: the FT policies for the domains (*FTFaultToleranceDomain*), the identification of element groups which need to be monitored with FT mechanisms (*FTServerObjectGroup*), the state to be considered in state-full-replicas (*FTLoggableState*

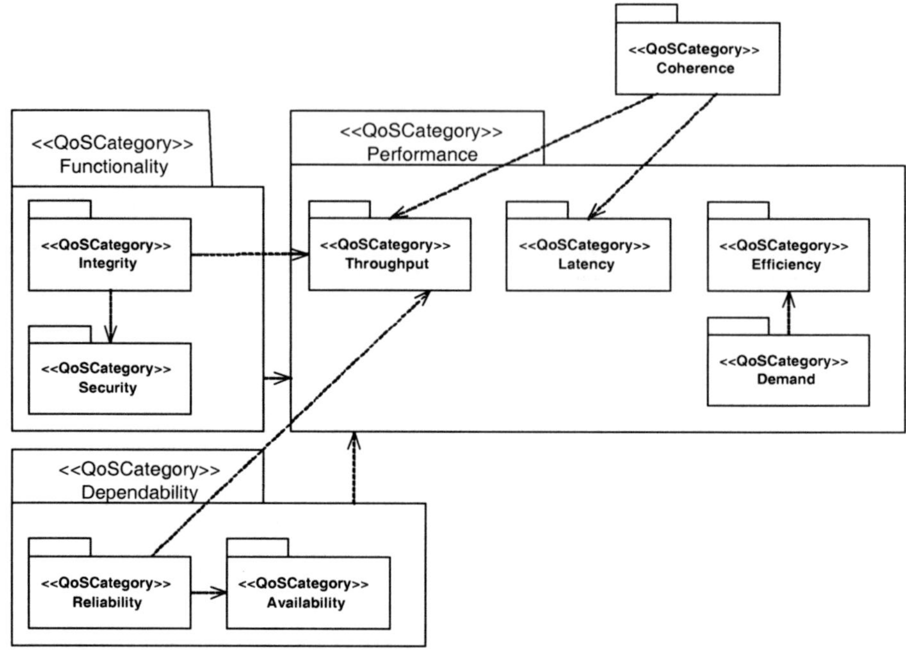

Fig. 7.25. Subprofiles of the QoS profile

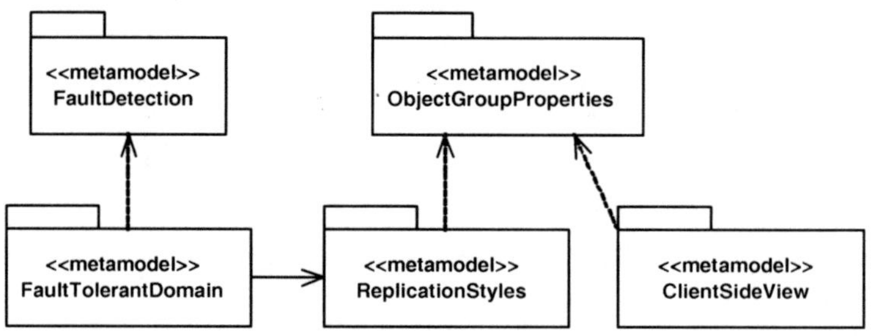

Fig. 7.26. Metamodel of FT architectures

and *FTHasReplicationState* with the replication styles *FTReplicationStyle* and subclasses). Figures 7.27 and 7.28 include the general stereotypes and further ones to describe replication styles. These stereotypes are based on metaclasses and attributes presented in Fig. 7.26.

7.4 Project Life-cycle Management in UML

Although being a methodology, UML does not prescribe life-cycle management by itself. Guidelines have been developed in the form of checklists (e.g. [6]), which

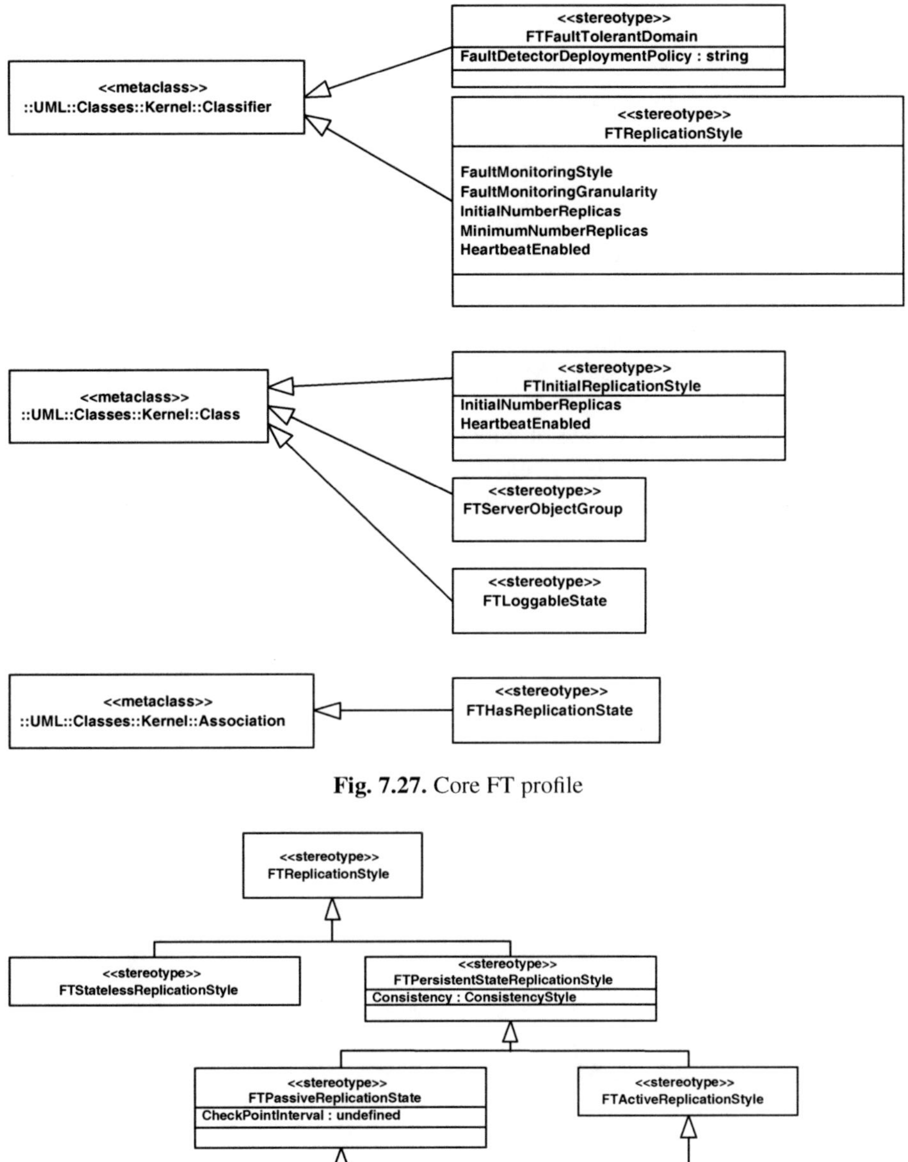

Fig. 7.27. Core FT profile

Fig. 7.28. Profile of FT replication styles

introduce intermediate checks after typical phases in the life-cycle of UML-based design projects:

1. Identification of use-cases, packages, classes (and capsules in UML-RT, and structured classes in UML 2.0)
2. Identification and specification of associations (ports in UML-RT and UML 2.0), and attributes
3. Definition of inheritance, associations, attributes (and protocols in UML-RT, and interfaces in UML 2.0)
4. Specification of sequences, state chart diagrams and operations

Figure 7.29 shows the typical life-cycle of UML projects. Some diagrams, which have been excluded/replaced in UML 2.0, have been intentionally left out, since they mainly represent joining diagrams for the listed basic diagrams and will be dealt with later. On the left-hand side of the picture, the structural constructive steps are listed, whereas on the right-hand side the behavioural constructive steps are shown in parallel. This means that the two sides can be devised simultaneously with minor correlations, producing a complete UML model at the end.

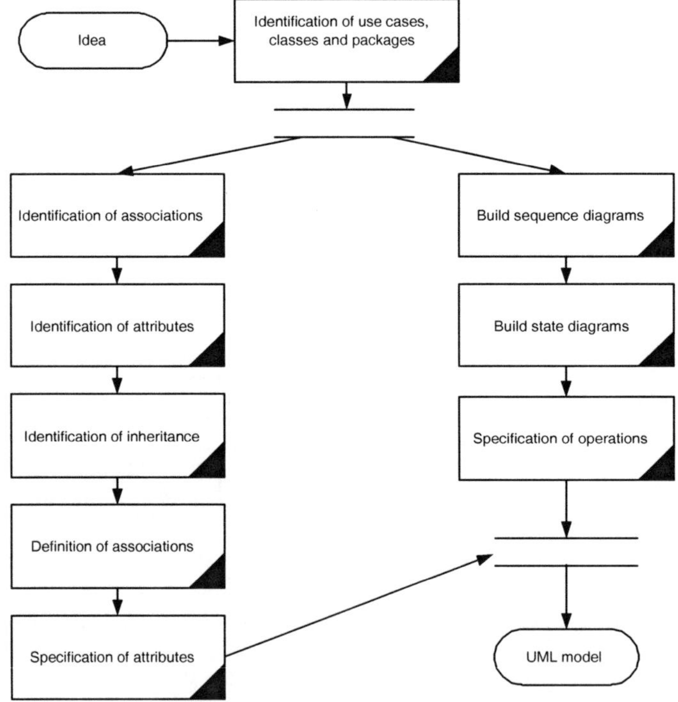

Fig. 7.29. UML project life-cycle

Moreover, one can introduce appropriate QoS measures into design processes and designed models in order to increase or ensure the QoS of systems designed. A systematic approach to doing this in standard UML is described in the profile for modelling QoS and fault tolerance characteristics and mechanisms [55].

7.5 Design Verification in UML with Checklists

According to [74], one may wish to introduce verification and validation not only at the product level but also at the design or model levels to ensure QoS of UML-based designs. Verification and validation of models are quality assurance techniques, which are meant to prevent as well as to detect design errors, being the possible sources of *faults, inconsistencies* and *incompleteness*. They comprise a set of activities, which should ensure that models are correct. Verification usually focuses on software correctness, whereas validation ensures that software meets its users' needs: *"Validation deals with tracing the software to the requirements"* [61]. In this sense, verification and validation checks and quality assurance techniques for design models are composed of *audits (aesthetics), reviews and inspections (syntax),* as well as *walkthroughs (semantics).*

In the subsequent sections the checklists pertaining to individual steps in the life-cycle (see Fig. 7.29) are presented. As in [6], they are formed of constructive steps followed by analytical steps, which appears to be the most efficient "waterfall" model of a life-cycle with intermediate checking. They have been supplemented for checking designs formulated in UML-RT as well as UML 2.0. The checks mainly relate to *clarity, completeness, parameter compatibility* and *comprehensiveness*, as well as to *well-formedness of operations*.

7.5.1 Use-case Checklist

Results:

Use-case diagram: considering the designed system as a black box, all use-cases and (external) actors are identified and included in the diagram.

Use-case specification: all use-cases are described in natural language or within a tabular scheme.

Construction:

Actor identification: persons performing certain operations with a (software) system, interfaces or other systems, which communicate with the system.

Operation identification: activities, which are carried out/triggered by actors and return results to them; for each externally triggered and time-dependent operation a use-case should be defined.

Standard activity: executed in more than 80% of all cases should be identified; in addition, the *ideal and minimum* use-cases should be identified.

Check:

Too complex? Is it possible to simplify complex use-cases? Complex activities can be broken down and specified as stand-alone use-cases using << *include* >> and << *extend* >> relations among them.

Well formed? A "good" description of a use-case is (1) easily understandable to the customer, (2) a professional description of the functionality, (3) a complete description of the normal behaviour, (4) contains a separate description of exceptional behaviour, and (5) is not longer than one page.

Non-redundant? Common activities are joined with << *include* >>.

Consistent? Are use-cases consistent with class diagrams? Test by building an object diagram.

Exceptions: enhancements and alternatives should be specified in schemes and use-case diagrams (e.g. using << *extend* >> relations).

Additions: activity diagrams can be used to show the logical order of events from/to actors and activities performed in a use-case. Alternative orders can be specified by introducing *decision nodes*. Also, parallel execution can be specified by introducing *fork/join* structures.

7.5.2 Package Checklist

Results:

Package diagrams: with the identified packages and their interdependencies being marked.

Packages: are assigned model elements.

Construction:

Top-down: Which packages are devised by the "top-down" approach? For big projects systems should be broken down into packages, representing subsystems, possibly even before specifying use-cases with the main packages potentially broken down into subpackages.

Bottom-up: Which packages are obtained by the "bottom-up" approach? Here, classes are integrated under a common denominator.

Check:

Consistency? Does a package represent a consistent whole? Classes belonging together are joined in the package, inheritance structures lie possibly within the package, aggregations are not broken, as few as possible associations are broken.

Name? Is the package name suitable? The package contents should be expressible by up to 25 words; its name should represent its descriptive title.

Size? Is the package too small? If it contains just one or two classes, it might be sensible to include it in another package or main diagram; comprehensibility is the main criterion here — if it is enhanced by the presence of the package or if its growth is expected, it should be sustained, otherwise not.

Exceptions: inheritance structures may be broken only horizontally — the classes in the package are inherited from the classes of a single package, only.

7.5.3 Class Checklist

Results:

Classes: with their most important properties and methods are added to the class diagram. For any class, whose name is not descriptive enough, a description of up to 25 words is added.

Capsules (UML-RT): are added with the same characteristic properties as classes have.

Construction:

Bottom-up: Which classes can be identified from specification documents by "bottom-up" analysis? Their analysis or re-engineering of old software are also the origin of class attributes.

Top-down: Which classes can be identified from the use-case descriptions by "top-down" analysis? Search description for substantives and their adjectives to discover classes with their properties and verbs to discover potential operations.

Category: To which categories do classes belong? This leads to the identification of possible packages or structured classes (UML 2.0).

Check:

Name? Is the class name descriptive enough? It should be taken out of professional terminology and have the same meaning as the whole of its attributes.

Abstraction level? Is the current level of abstraction accurate? Should the class be broken down into multiple classes and represent a package?

Size? Are the classes too small? Can they be expressed as attributes of another class?

Number? Are the classes too many? Is grouping classes into packages considered?

Inappropriate design? Do classes model screen patterns? Do classes model development/implementation details?

Exceptions: do not construct classes just to manage groups of objects.

Additions: Capsules/Structured classes? Do classes represent active objects? UML-RT introduces capsules instead of classes for these objects. An active object also communicates with other objects — identify other active objects and assign ports to both. Connect them with protocols. UML 2.0 introduces structured classes instead of capsules and assigns them ports as well. Define provided and required interfaces.

7.5.4 Protocol Checklist

Results:

Protocols: are defined in UML RT. They encompass incoming and outgoing signals. Capsule ports are interconnected through protocols by establishing aggregation connections among capsules and protocols.

Interfaces: are defined for ports of interconnected classes in UML 2.0. Protocols are defined by provided and required interfaces for any port.

Construction:

Identify signals/messages: Which data are exchanged between active objects? The identification of data and their types leads to the definition of incoming and outgoing signals/messages.

Identify interfaces: Which methods are invoked from other active objects? The identification of messages invoking execution at another object leads to trigger signals which must also be accounted for in the corresponding state diagrams of classes/capsules.

Check:

Consistency? Are interconnected ports compatible? One's input signals should be another's output signals (UML-RT). In other words (UML 2.0), one's required interface should represent a subset of another's provided interface.

Interfaces chosen appropriately? Do not draw ports just to model method invocations (these are modelled by associations).

7.5.5 Association Checklist

Results: associations should be established during identification only, by drawing lines (their cardinalities, aggregations, compositions, rôles, names and restrictions should be added while defining them).

Construction:

Which associations can be devised from document analysis?

Which associations can be devised from use-cases? Look for verbs describing physical nearness/containment, actions, communications, properties or relations.

Which restrictions does the association have to fulfil? For instance, if we wish to stress that the objects of the class from the other end of a single association have to be ordered correspondingly for the other side to process them correctly, we may do so by writing the association restriction {ordered} (we might also include the ordering strategy, e.g. {ordered, FIFO}). In the case of multiple associations, linking a class with other classes, we may wish to stress that at least one of them or a subset thereof needs to be established among the corresponding objects at run-time by specifying association restrictions {or} or {subset} with the associations.

Which rôles do the classes they connect perform? Assign rôle names when (1) associations exist between objects of the same class, (2) a class plays a different rôle in different associations or (3) the rôle name better describes the class/association.

Check:

Rôle name necessary/appropriate? A rôle name is necessary when more associations exist between two classes. Rôle names are preferred as opposed to association names.

"1:1" association? Two classes should only be modelled when (1) the connection is optional either way, (2) two big classes are connected or (3) if their semantics are different.

Multiple association? Are there two or more associations between two classes? Do they have different meanings/cardinalities?

Inheritance? If an association represents inheritance, do not include it.

Exceptions: Are inherited associations used correctly? Do not re-establish them, since they provide no additional information to the model.

7.5.6 Attributes Checklist

Results:

Attributes: are simply assigned to classes by their names in the first iteration.

Attribute descriptions: in the second iteration the individual attributes are completely described (type, access, initial value) and complex types are defined (possibly introducing new classes).

Construction:

Derive attributes from specification documents (elementary data can easily be grouped into data structures).

Derive attributes from use-cases (which data are necessary to perform a use-case, or are used in decisions of the activity diagram, respectively).

Check:

Coherency? Have the correct attributes been chosen? Check their relevance and whether they change in time.

Type? Were the attribute types chosen correctly? Introduce additional types in case base types are not sufficient.

Names? Are attribute names appropriate? They should be short, nouns or noun adjectives that do not replicate the class names. Only generally accepted acronyms shall be used.

Abstraction level? Do attributes belong to the same abstraction level? Only then shall they belong to classes of the same level.

Class:Association? Does an attribute belong to a class or an association? Does it pertain to the class if observed separately?

Inheritance? Are inherited attributes used correctly? They should enhance readability/understandability.

Exceptions: check attributes for class capability. Classes should be introduced only if they possess object identities.

7.5.7 Inheritance Checklist

Results: inheritance structures are drawn in the diagram. At their origin abstract classes are formed.

Construction:

Generalisation: "Bottom-up" — do similar classes exist for which it is possible to define a general class?

Specialisation: "Top-down" — can each object of the class be assigned any attribute value, can it execute any of its operations?

Check:

Is-a? Is there an "is-a" relation in the model? They are candidates for inheritance structures.

Meaningful? Does the inheritance structure increase the comprehensibility of the model? If not, it should be omitted, and the corresponding classes joined.

Consistency? Does every inherited class use all inherited attributes and operations? If not, an alternative abstract class should be constructed containing only attributes from the intersection of the derived classes.

Natural? Is the inheritance structure "natural" in terms of the system being modelled (in terms of the system model)?

Number? Are there at most three to five inheritance levels? More levels decrease the comprehensibility of the model.

Exceptions: subclasses do not require inheritance if they only represent different kinds of a class which do not differ in their attributes nor operations. In this case they probably differ just in attribute values.

7.5.8 Cardinalities Checklist

Results: cardinalities are assigned to (edges of) associations in a class diagram representing numbers of class instances on respective sides thereof.

Construction:

Snapshot vs. history: Is a snapshot ("0" or "1" cardinality) sufficient or is a history (cardinalities "n" or "*") to be modelled? While with a snapshot an "old" connection is relinquished when a new one is established, with a history a new one is added between the active objects.

Shall vs. must: Is there a "shall" (cardinality lower bound $= 0$) or "must" (cardinality lower bound ≥ 1) association? With a "must" association the two objects need to co-exist in time with their association established as soon as the "must"-side object becomes active.

Check:

Coherency? Was a "must" association used, where it was not needed? If we do not require history, usually a "shall" association is sufficient.

Consistency? Does the cardinality contain concrete values? The upper bound can be defined by the problem domain.

7.5.9 Aggregation Composition Checklist

Results: aggregation and composition structures are drawn into the class diagram.

Construction:

Composition: represents a "part-of" relation; the life-cycle of the parts is related to the one of the whole; the function of the whole is transferred to its parts (patterns like component lists are a good orientation).

Aggregation: represents a "part-of" relation, where a part object may be part of multiple aggregate objects.

Composition check: Would modelling (the attributes of) part classes as attributes of a composite class lead to a bad model? If so, composition is justified.

Exceptions:

Aggregation is rare.

When in dilemma, always use "simple association".

7.5.10 Sequence Diagram Checklist

Results: sequence diagram is built for every relevant path through a use-case. Alternatively, a collaboration or communication (UML 2.0) diagram may be used.

Construction:

From any use-case, scenarios can be derived in the form of sequence and collaboration diagrams. Normal-state and exception-state execution scenarios should be identified, classified and rated by their importance. Since they lead to different responses/results, a sequence or collaboration diagram should be built for each of the relevant diverse scenarios.

Derive the classes involved. They represent active objects (capsules in UML-RT, or structured classes in UML 2.0).

Break up tasks into operations. They are candidates for class operations and messages exchanged through *capsule ports* in UML-RT and interface operations of *structured classes* in UML 2.0.

Check the sequence of operations and order them accordingly in the diagram.

Assign operations to classes depending on the attributes involved.

Check:

Lifetime? Are the recipient objects reachable? In the case of a mandatory ("must") association this is statically fulfilled, with optional ("shall") associations this must be dynamically fulfilled by the class operations.

Consistency? Is the sequence diagram consistent with the class diagram? All classes are also present in the class diagram, and only operations defined there are also used in the sequence/collaboration diagram.

Exceptions: caution has to be taken in order not to describe the user interface instead of the functionality, or to go into too much detail (describing operations).

Additions: in UML 2.0 *activity diagrams* have been substituted by *interaction overview diagrams*, where collaboration and sequence diagrams can be combined on the project level providing better overview and comprehensibility.

7.5.11 State Diagram Checklist

Results: state diagrams should be built for any non-trivial life-cycle of an object.

Construction:

A non-trivial life-cycle exists, if the same event can trigger different behaviours depending on the object's current state.

The states are determined by the start state and depend on the events and subsequent states. They are defined by attribute values and connections between objects.

Check:

State name? It should be a verb in passive voice, at best representing the situation/activity.

End state exists? If an object seizes to exist or becomes inactive by the end of its life-cycle.

Consistency? Check operation compatibility with the class. In the state diagram only the state-dependent operations of the class shall be included.

Activity vs. action? Are operations to be modelled as activities or actions? Activities represent in-state operations and are bound to them, while actions represent atomic (transition) actions.

Events? Which events are to be modelled? Modelled events are ones from the user, from other objects, periodic events, timers, internally generated ones (signals), exceptions, and errors.

State transfers? Are all state transfers correctly described? Are all states reachable? Can any state be left, with the exception of the end state? Are state transfers unique (check transition conditions)?

Exceptions: beware of loops — transition conditions must always be associated with events.

Additions:

In UML RT a state chart has to be defined for any capsule to model its life-cycle and handle its communication with associated capsules.

In UML 2.0 timing diagrams have been introduced as links between sequence and state diagrams to represent associated execution scenarios and value changes of active objects. Common (trigger) messages introduce both.

7.5.12 Operations Checklist

Results:

Operations: are added to the classes of the class diagram.

Descriptions: are added for operations whose names are not descriptive enough.

Construction:

Operations are derived from use-case scenarios, sequence/collaboration and state diagrams.

Consider operation inheritance — place operations as high in the hierarchy as possible.

Add descriptions (comments) to operations using natural or semi-formal language.

Check:

Coherency? Does an operation fulfil the necessary quality criteria? Is it not too extensive, does it fulfil functional binding criteria (i.e. one function for each operation)?

Balancing? Is balancing sufficient? Classes should possess attributes needed by their operations or their descendant operations only.

Exceptions: care must be taken not to model user dialogues instead of pure functionality.

Additions: in UML 2.0 operations acting as interfaces have to fulfil the additional functionality required by the protocol (accept/submit signals/messages).

7.6 Concluding Remarks on UML-oriented Design for QoS

UML methodology, especially in its second version, has reached a level of maturity, which does not only provide it with the means to design a large variety of information systems including real-time systems, but also with the ability of doing this in a QoS-oriented way. It enables the inclusion of timing analysis as well as fault tolerance mechanisms into designs. One can check designs for completeness as well as parameter compatibility, which is as far as one can get on the model level to ensure model correctness and the QoS level requested. As can be seen from the checklists, a kind of life-cycle management of UML projects has already been envisaged. Since it is not part of the standards, projects cannot be certified in accordance with it. Further design and development efforts towards certification of design models are described in the following chapter.

8

Certification of Real-time Systems

Over the past decade, design engineers have been forced to examine vendors' quality issues more closely. Mostly hardware has been under consideration, but now, due to the increasing complexity of real-time systems, the same rigid quality constraints must be applied to both the hardware and software selection process. Some examples of both hardware and software standards are given in the subsequent section. Cross certification between hardware and software suppliers is of paramount importance, since system reliability is largely determined by the interaction between components, and not just the individual components. More and more designers today rely on the standard ISO 9000 to ensure that vendors have robust quality programmes in place [22]. An overview of the ISO 900X standards is given later in this chapter. Finally, some technical evaluation standards that describe the way to use standards to evaluate hardware/software systems and provide them with certificates when they comply with the requirements specified are considered.

8.1 Technical Standards

8.1.1 Telecommunication Standards

Being genuine real-time systems, telecommunication systems and their standards have strongly influenced the development of other fields of real-time computing. Hence, their standards are among the first and the most comprehensive and well defined ones today. Standards are needed for reasons of interoperability, quality assurance and consistency of their specific field's evolution. There are three different kinds of organisations active in standardising telecommunication systems:

1. International or governmental (e.g. ITU)
2. Semi-official (e.g. ETSI)
3. Volunteer organisations (e.g. 3GPP or 3GGP2)

The biggest official standardisation organisation in the field of telecommunications is the International Telecommunication Union (ITU), which is further subdivided into the following three sectors:

1. Radio communication (R)
2. Telecommunication (T)
3. Telecommunication development (D)

The European Telecommunication Standardisation Institute (ETSI) was founded in 1988 and co-operates closely with ITU.

The most important telecommunication standards are available from the following sources:

3GPP (http://www.3gpp.org) The 3rd Generation Partnership Project (3GPP) is a collaboration agreement that was established in December 1998. Originally, 3GPP was meant to produce globally applicable technical specifications and technical reports for a third generation mobile communication system based on the core networks of the Global System for Mobile communication (GSM) that had evolved and the radio access technologies that they support (i.e. Universal Terrestrial Radio Access (UTRA) for both Frequency Division Duplex (FDD) and Time Division Duplex (TDD) modes). The scope was subsequently enhanced to include maintenance and development of the GSM technical specifications and technical reports including evolved radio access technologies (e.g. General Packet Radio Service (GPRS) and Enhanced Data rates for GSM Evolution (EDGE)).

Bluetooth (http://www.bluetooth.com) is a wireless technology for short-range communication intended to replace cables connecting portable and/or fixed devices while maintaining high levels of security. The Bluetooth specification provides product developers with definitions both for the link and the application layers, which support data and voice applications. Its main benefits are robustness, low energy consumption and low cost.

DECT (http://www.dectweb.com) Digital Enhanced Cordless Telecommunication is a technology adopted worldwide for cordless telephones, wireless offices, and wireless telephone lines to the home. In contrast to GSM, it is a radio access technology rather than a comprehensive system architecture. DECT has been designed and specified to interoperate with many other types of networks, such as PSTN (conventional telephone networks), ISDN (digital and data telephone networks) or GSM (mobile telephone networks).

IEEE 802(R) (http://standards.ieee.org/getieee802) This family of wireless communication standards includes 802.20 on Mobile Broadband Wireless, 802.22 on Wireless Regional Area Networks (WRAN), and 802.11 on Wireles Local Area Networks (WLAN/Wi-Fi).

TETRA (http://www.tetramou.com) "TErrestrial Trunked RAdio" comprises a suite of open standards for digital trunked radio defined by ETSI to meet the needs of the most demanding users of Professional Mobile Radio (PMR). TETRA is an interoperability standard for equipment from different vendors. It is used by PMR users in the areas public safety, transportation, utilities, government, commerce, industry, oil and gas and military. TETRA is also used by operators of Public Access Mobile Radio (PAMR).

Open Mobile Alliance (OMA) (http://www.openmobilealliance.org/tech/affiliates/wap/wapindex.html) Formerly known as Wireless Application Protocol (WAP) Forum, OMA provides application level standards (e.g. format, security) to transfer data from the Internet and visualise them on mobile devices.

Among the standards specifically defined for automotive and industrial communication are:

Profibus (http://www.profibus.com) There exist three basic versions: (1) Profibus-FMS (Fieldbus Message Specification) to communicate between automation devices based on the client-server model, (2) Profibus-DP (Decentralised Peripherals) to connect sensors and actuators to controlling devices for fast remote input/output, and (3) Profibus-PA (Process Automation) to connect field devices and transmitters to process control devices such as inverters on SCADA networks. The fieldbus allows for intrinsic safe transmission and power on the line. Device/network parameters and *function blocks* have been defined to cover the need of process engineering. Profiles that contain additional functionality for safety applications (PROFIsafe), for motion control (PROFIdrive) and other types of devices are available. Moreover, a light-weight version for small Ethernet-based networks (ProfiNet) has benn defined. Profibus was standardised 1991/1993 as DIN 19245 in Germany, 1996 as EN 50170 in Europe and 1999 as IEC 61158/IEC 61784 worldwide.

CAN (http://www.can-cia.com) The Controller Area Network is a broadcast, differential serial bus, originally developed in the 1980s by Robert Bosch GmbH, to connect electronic control units (ECUs). CAN was specifically designed for robust communication in electromagnetically noisy environments. It can use a differentially balanced line like RS-485 as medium. Employing twisted pair wires it becomes even more robust against noise. Although initially created for the automotive industry (vehicle bus), nowadays it is used in many embedded industrial control applications. Its messages are small (8 data bytes maximum) and are protected by a CRC-15 (polynomial 0X62CC) that can guarantee a Hamming bit length of 6 (so up to 5 corrupted bits in a row will be detected by any node on the bus). The CAN data link layer protocol was standardised in ISO 11898-1 (2003), where the data link layer, composed of the Logical Link Control (LLC) and Media Access Control (MAC) sublayers and some aspects of the physical layer according to the OSI Reference Model are defined.

Profibus and CAN are the most representative fieldbuses considering their market shares. Their main properties, in addition to the ones of the previously mentioned telecommunication standards, are the emphasised safety and robustness of communication. There exist multiple derivatives of both, which have been developed for time-triggered applications (e.g. Flexray based on CAN or RT-Ethernet). None of them can, however, guarantee temporally predictable communication or application execution.

The standards pertaining to telecommunication systems address the entire development cycle, from planning, design of hardware and software platforms, via development, to testing and maintenance. They were also the first to introduce certificates for standard-compliant devices. So, users know exactly the level of service they can expect from purchased equipment, and its producers have a good reference on the minimum required performance and functionality.

8.1.2 Operating Systems Standards

Besides reliable hardware and robust communication, the next critical corner-stones of dependable application execution are operating systems. The reasons why standard operating systems and corresponding benchmarks, which have proven to render relevant citable results, are not always used have to do with pragmatic decisions on their suitability for specific applications. Since the answers to questions such as does an operating system provide the functionality required, is its performance acceptable or does a commercially available implementation exist for the hardware platform chosen are not always affirmative; often non-standard implementations are used and, hence, more general benchmark solutions have been striven for.

POSIX in Real Time

A good example of standardisation with respect to real-time operating systems is POSIX. While the original Portable Operating System Interface for Computing Environments was published in 1990 as POSIX 1003.1 [29], subsequent releases of POSIX have been defined to cover real-time extensions and multithreading (POSIX 1003.1d, 1003.1j), distributed systems (POSIX 1003.21) and high-availability services (POSIX 1003.2h) as well.

OSEK/VDX Standard

Another example of an operating system standardised for real-time systems is OSEK/VDX (www.osek-vdx.org) [67]. It defines the operating system itself (OS part) with its communication (COM) and network manager (NM) parts. Initiated by the automotive industry, this operating system has turned out to be useful for small embedded systems as well as for larger distributed and networked ones. It is also considered an industry "benchmark" for automotive electronics. Characteristic for the operating system standards is that they define different levels of conformity, based on the use of their functions or their parts. So, one may say that the level of conformity somehow characterises the application with respect to standard compliance.

8.2 Technological Standards

8.2.1 ISO 900X Standards

An organisation can be certified to comply with one of the four levels according to ISO 9000, with ISO 9001 being the most complete:

ISO 9001: model for quality assurance in design/development, production, installation, and service
ISO 9002: model for quality assurance in production and installation
ISO 9003: model for quality assurance in final inspection and test
ISO 9004: quality management and quality system elements

Designers of embedded and real-time systems can buy both hardware and software building blocks or bundles thereof from suppliers certified according to ISO 9001. The use of high-quality off-the-shelf components gives them a head start in reducing development time and cost of their applications. This also simplifies their ISO 9001 certification process at the level of system integration, since certified components are simply included in their own quality considerations.

8.2.2 Capability Maturity Model for Software

There is no standard integration of the ISO 9000 standards into any software engineering method, not even into UML. The closest form of their "coupling" so far is the Capability Maturity Model (CMM) for Software [59], a software engineering process that has to fulfil nearly the same requirements as prescribed by the standard ISO 9001. Hence, to integrate the two, the UML project life-cycle should be adjusted in a way that supports CMM.

Software Process Maturity is defined by CMM as the extent to which a specific process is explicitly defined, managed, measured, controlled and effective. Maturity implies a potential for growth in capability and indicates both the richness of an organisation's software process and the consistency with which it is applied in projects throughout the organisation. As an organisation gains in software process maturity, it institutionalises its software process via policies, standards and organisational structures. Institutionalisation entails building infrastructures and corporate "cultures" that support the methods, practices and procedures of the business so that they endure after those who originally defined them have gone.

As shown in Fig. 8.1, CMM provides a framework to organise these evolutionary steps into five maturity levels that lay successive foundations for continuous process improvement. These five maturity levels define an ordinal scale to measure the maturity of an organisation's software process and to evaluate its software process capability. The levels also help an organisation to prioritise its improvement efforts. A maturity level is a well defined evolutionary foundation towards achieving a mature software process. Each maturity level (see Fig. 8.2) comprises a set of process goals that, when satisfied, stabilise an important component of the software process.

Achievement of each level of the maturity framework establishes a different component in the software process, resulting in an increase in the process capability of an organisation:

1. At the *Initial Level*, organisations do not provide stable environments to develop and maintain software. Their competence rests on capable individuals and success is entirely due to them.

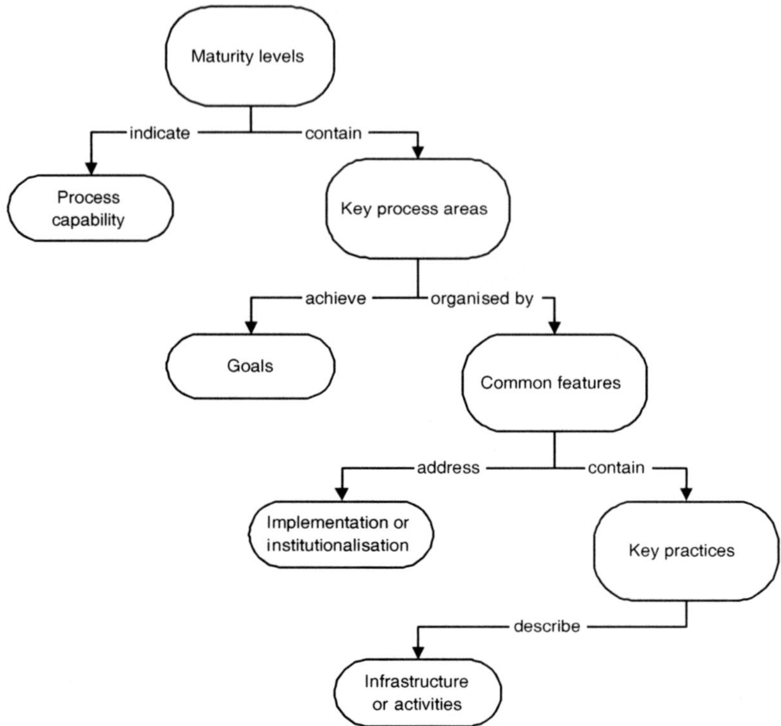

Fig. 8.1. CMM structure

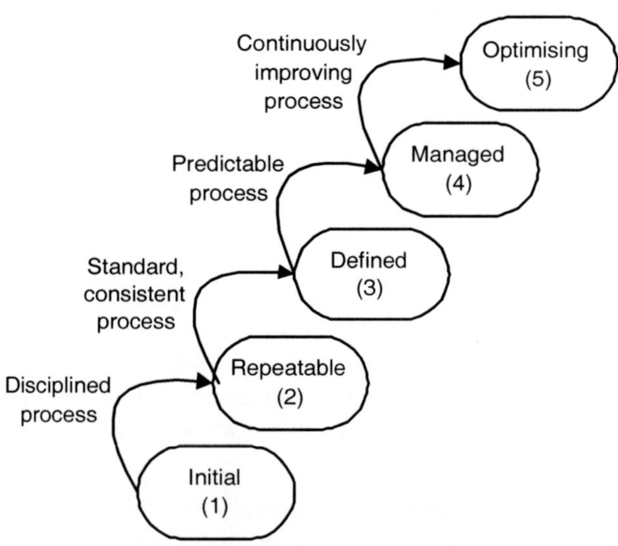

Fig. 8.2. Maturity levels of CMM

2. At the *Repeatable Level*, organisations have developed and implemented traceability into their project management processes — projects are conducted based on experiences gained in similar previous ones. Usually, at the end, the results are evaluated, opening opportunities for process improvements.
3. At the *Defined Level*, the standard processes to develop and maintain software across organisations are documented, including both software engineering and management processes, which are integrated into coherent wholes. These standard processes are referred to throughout the CMM as the organisations' standard software processes. The software process capability of level 3 organisations can be summarised as standard and consistent, because both software engineering and management activities are stable and repeatable.
4. At the *Managed Level*, organisations set quantitative quality goals for both software products and processes. Productivity and quality are measured for important software process activities across all projects as part of organisational measurement programmes. The software process capability of level 4 organisations can be summarised as quantifiable and predictable, because their processes are measured and operate within measurable limits. This level of process capability allows organisations to predict trends in process and product quality within these quantitative bounds.
5. At the *Optimising Level*, entire organisations are focused on continuous process improvement. They have the means to identify weaknesses and strengthen their processes proactively, with the goal of preventing defect occurrences. Software project teams in level 5 organisations analyse defects to determine their causes.

Each maturity level is a composition of its founding parts. With the exception of level 1, the decomposition of each maturity level ranges from abstract summaries of each level down to their operational definition in terms of key practices, as shown in Fig. 8.1. Each maturity level is composed of several key process areas (see Fig. 8.3). Each one of the latter is organised into five sections called common features, which specify the key practices that, when collectively addressed, accomplish the goals of the key process area. More details of the key process areas and practices can be found in e.g. [59, 60].

8.2.3 BS 7799 Security Standard

Design for security encompasses a comprehensive list of rules, regulations, code of good practice and mechanisms (see Sect. 6.2). This is also how the standards pertaining to security like the BS 7799 security guide were introduced. It first appeared in 1995 as a re-publication of the British Department of Trade and Industry's (DTI) "Information Technology — Code of practice for information security management", was upgraded in 1999 and became the ISO 17799 [34] standard in 2000. It was further developed in 2002 as BS 7799-2, focusing on implementation of information security management systems and became ISO 27001 in 2007 [35]. After a final revision in 2007, ISO 17799 was re-named ISO 27002 [36]. Its latest addition BS 7799-3, published in 2005, covers risk analysis and management. It aligns with ISO 27001.

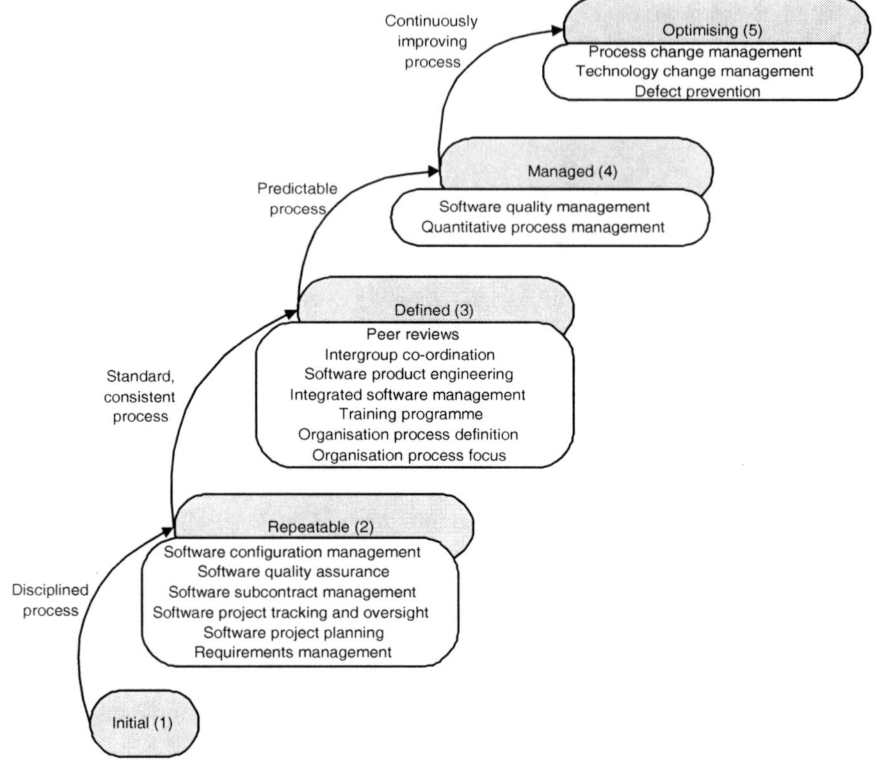

Fig. 8.3. Key process areas by maturity level

Any company striving for a security certificate should comply with the mentioned (and related national, e.g. [50]) standards' regulations. The quality management systems of companies holding ISO 9001 certificates can easily be upgraded with this security management system, since the standards have been aligned.

8.3 Technical Evaluation Standards

8.3.1 Evaluation of Software Products according to DIN 66272 / ISO/IEC 9126

The German standard DIN 66272 [15], which incorporates ISO/IEC 9126, defines six properties characterising software quality. They form the foundation for specialisation and more specific descriptions of software quality:

Functionality: a set of properties relating to the presence of functions and their pre-defined properties that fulfil pre-defined requirements

Dependability: a set of properties characterising the ability of software to fulfil and sustain required operating conditions over a specified time-frame

Usability: a set of properties relating to the effort required for use; should be evaluated individually with the target users

Efficiency: a set of properties relating to the proportions between performance of software and amount of required resources under pre-defined conditions

Changeability: a set of properties relating to the effort required to perform pre-defined changes

Portability: a set of properties relating to the suitability of software to switch from one environment to another

The definition of the characteristics of software quality and the quality review model above are applicable to software products and quality audits during their life-cycle. The standard also defines quality requirements for software and can be used for its evaluation (measurement, ranging and judgement). To be able to do this it is necessary to:

- Define the quality requirements for a software product
- Evaluate the software specification to determine whether it fulfils the quality requirements of the development process
- Describe the indicators and properties of the software realised (e.g. in user manuals)
- Evaluate the realised software before deployment
- Evaluate the realised software before being taken over by the customers

Generally accepted metrics for the mentioned quality properties are rare. Hence, standardisation and evaluation groups are left the freedom to develop their own validation methods and metrics for different application areas and phases in the development cycle even by "rules of thumb". To consider the six quality criteria, it is necessary to introduce leverage and criteria that correspond with organisation and application area.

There are three different points of view on software product quality: the user's, the developer's and the manager's view. They differ slightly. The user is mainly concerned with functionality, dependability, efficiency, ease of use and portability, whereas besides these the developer is also concerned with the quality of partial results and phases of the development process. The manager, on the other hand, is more concerned with the overall impression of the product's quality and quality improvement, while sustaining the overall project's time and cost margins.

The evaluation process of software quality is, as shown in Fig. 8.4, based on the mentioned six properties of software products. It consists of three phases: setting up quality requirements, defining evaluation and proper evaluation. The process can be applied to any software component. The result of the evaluation step summarises the measurements of quality evaluation (see Fig. 8.5). The end result is a statement about the quality of a software product. Is is compared to other significant indicators such as time or cost, followed by the management's decision on the software product's (non-) release.

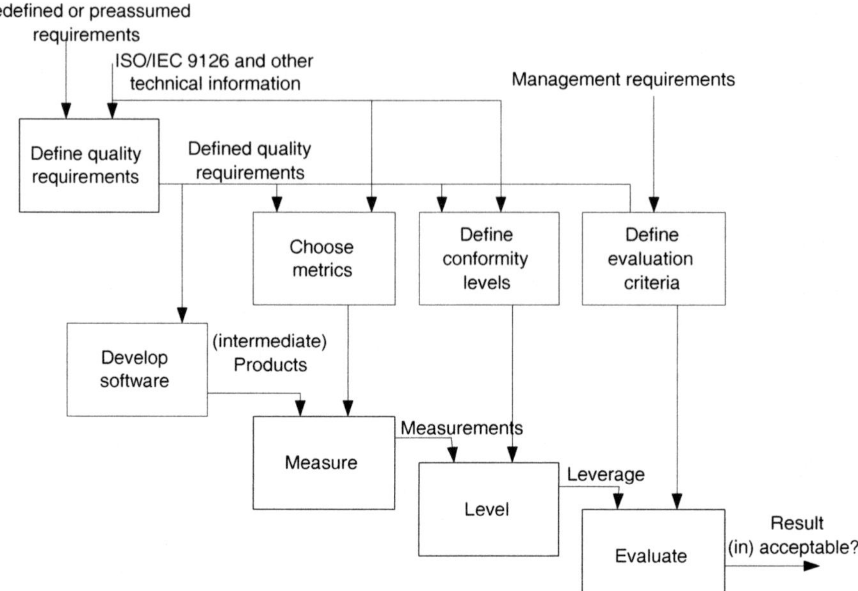

Fig. 8.4. Evaluation process model

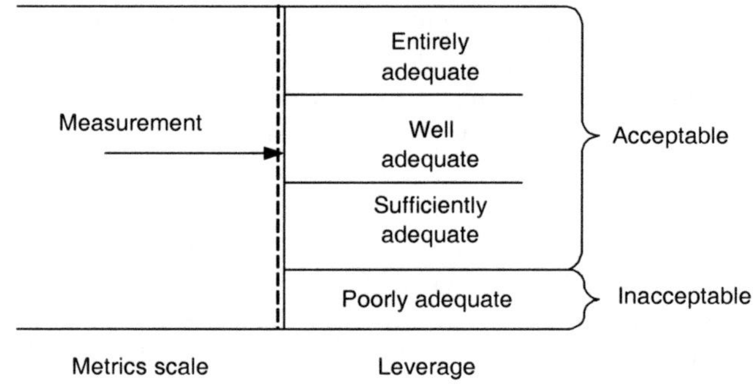

Fig. 8.5. Quality measurements levels

8.3.2 Measuring and Rating Data Processing Performance with DIN 66273

This standard [16] is equally important for software providers as for their clients. It is meant to define standard-compliant requirements as well as to make statements on data processing performance. With its help the performance of data processing systems or their components can be uniquely described. It prescribes:

- Terms and metrics
- Measurement methods
- Methods to produce working profiles in form of models
- Performance evaluation

The standard describes a method for performance assessment, which leads to objective performance indicators. The measured objects are considered by their evaluators in the same way as by their users, namely as "black boxes". Performance indicators of internal components are not important for overall performance statements. For bottleneck analysis, it is sometimes necessary that the performance indicators of different components are recorded simultaneously. The standard does not make any statements on that. It foresees operation with a synthetic workload. This requires a driver software, which creates working profiles for the data processing systems investigated. The assumption is made that the execution times of the operations are recorded automatically.

The standard defines vectors of performance indicators, which represent utilisations (b_j) and mean execution times (t_{AN_j}). From the throughput times measured in operation the relative frequency of operations, whose $t_{A_j}(k)$ values do not exceed a reference value t, is obtained. A typical description of operation requirements specifically states the allowed deviations of the throughput time values and, herewith, determines the required performance. Usually, operations are allotted places in operation urgency classes with upper limits $g_{F_j}(k)$ on throughput times for each of z_j operation classes. The relative frequencies of operations performing in the individual time limit classes are determined by $r_{F_j}(k)$.

Performance indicators are defined to evaluate performance. These are reference utilisation B_{Ref_j}, reference mean throughput time value t_{Ref_j}, time limit classes $g_{Ref_j}(k)$, and their relative frequencies $r_{F_j}(k)$. To obtain these values, real or theoretical reference systems can be defined with which the system under evaluation is compared. Preferably the measurement profiles obtained from real reference systems are used. Although optimal, the results obtained this way typically cannot be reproduced. Hence, evaluation using standard benchmark driver software is favoured by DIN 66273 [16]. Measurement methods are equal for all types of operations. The standard contains examples of programs in C, FORTRAN 77 and PASCAL for calculating the mentioned vectors.

To calculate system performance it is necessary to know the number of operations j executed in the time foreseen for each time class k, i.e. $e_{K_j}(k)$. The number of successfully executed operations is given by the sum $e_j = \sum_{k=1}^{z_j} e_{K_j}(k)$. Hence, the performance vector for all operations E is obtained by $E_j = \frac{e_j}{t_{measurement}}$. The number of vector components depends on the differentiation grade desired for a system.

8.4 Concluding Remarks on Certifying Real-time Systems

Standards explicitly addressing real-time systems are few, but the standards applying to telecommunication and data processing systems in general can be applied, and the standard mentioned last clearly shows how. The argument is, however, about measurement methods and benchmarks, which are, except for telecommunication systems, in general non-standard and, thus, rather subjective. So, a global standardisation organisation should come up with specific standards for developing real-time/embedded systems also for other fields. Crucial in this respect is a consensus on functions and performance of devices in a specific field of use. While some standards have already emerged (e.g. for the automotive industry), due to the broad diversity of applications this consensus has not yet been reached for all fields.

The final objective is, of course, the availability of standards for (real-time) software development, which would foster the improvement of their quality and performance. Appropriate evaluation standards in the sense of the most important system properties, among which for real-time systems the most significant ones are execution times and their variations under different working conditions, are under development.

To produce software systems from standard components is only possible if the mentioned standards are obeyed during their whole life-cycle. This requires software companies to reach at least CMM level 3. In addition, standard driver software (benchmarks) in the sense of DIN 66273 ought to be defined in order to enable standard-compliant evaluation of the products' performance.

These are the goals of the real-time community that, until today, have remained wishful thinking despite the mentioned (standardisation) efforts.

9

Conclusion

QoS measures for real-time systems have not received much attention, yet. Assessment of the few existing benchmark methods for real-time systems showed that they are not appropriate to evaluate overall QoS criteria.

A thorough look into the fundamental issues of real-time systems led to the proposal of a number of qualitative exclusive criteria, qualitative gradual criteria and quantitative criteria. Furthermore, it turned out that for hard real-time systems qualitative characteristics are much more important than quantitative ones. Provided that they fulfil the exclusive criteria, systems may then be compared on the basis of the gradual and quantitative ones. Since the latter mainly relate to costs, QoS comparison becomes possible — always under the constraint, however, that the indispensable exclusive criteria are met.

It was the purpose of this book to outline the state of the art, and to initiate a more thorough approach to the evaluation of QoS issues of real-time systems. The criteria proposed are to be regarded as guidelines, and one may sensibly seek compliance to further standards. They are not, however, standards themselves and cannot provide guarantees. Hence, for standard compliance, one must seek the aid of a corresponding authority. This material will be useful, however, as a reference to any developer or company to take proper precautions and to make the right decisions at the right times in the life-cycles of real-time projects.

References

1. Ada Performance Issues. *Ada Letters* 10(3), Winter 1990. SIGAda, ACM Press.
2. The Independent Agent's Guide to Systems Security — What Every Agency Principal Needs to Know. An Agents Council for Technology (www.independentagent.com/act) Report. http://na.iiaa.org/ACTDownloads/ACT1031505.doc
3. A. Andress: *Surviving Security — How to integrate People, Process and Technology.* New York: Auerbach 2004.
4. atp-Gespräch: Leistungstest für Prozeßrechner nach DIN 19 242. *Automatisierungstechnische Praxis* 32, 6, 315–319, 1990.
5. A. Avizienis, J.-C. Laprie and B. Randell: *Fundamental Concepts of Dependability*, Research Report N01145, LAAS-CNRS, April 2001.
6. H. Balzert: *UML kompakt: mit Checklisten.* Heidelberg: Spektrum 2001.
7. S. Behnsen: Betriebsystem für Embedded Applikationen: Minimierter Echtzeit-Posix-Kernel. *Systeme* 6, 36–40, 1996.
8. D. Brodarac: Was leistet ein Low-Cost-Personal-Computer im Vergleich zu einer microVAX II. *Automatisierungstechnische Praxis* 31, 5, 230–233, 1989.
9. BS 7799 and ISO/IEC 17799: Information Technology — Code of Practice for Information Security Management, 2000.
10. K.C. Cain, J.A. Torres and R.T. Williams: RT_STAP: Real-Time Space-Time Adaptive Processing Benchmark. MITRE Technical Report MTR 96B0000021, 1997.
11. H.J. Curnow and B.A. Wichmann: A Synthetic Benchmark. *Computer Journal* 19, 1, 43–49, 1976.
12. DIN 19 242 Teil 1 – Teil 14: Leistungstest von Prozeßrechensystemen: Zeitmessungen, 1987.
13. DIN 19 243 Teil 1 – Teil 4: Grundfunktionen der prozeßrechnergestützten Automatisierung, 1987.
14. DIN 44 300: Informationsverarbeitung No. 9.2.11, 1985.
15. DIN 66 272: Bewerten von Softwareprodukten – Qualitätsmerkmale und Leitfaden zu ihrer Verwendung, 1994.

16. DIN 66 273 Teil 1: Messung und Bewertung der Leistung von DV-Systemen – Meß- und Bewertungsverfahren, 1991.
17. B.P. Douglass: *Doing Hard Time: Developing Real-Time Systems with UML, Objects, Frameworks and Patterns.* Boston: Addison-Wesley 1999.
18. EEMBC Real-Time Benchmarks. http://www.eembc.com.
19. M. Eisenring, M. Platzner and L. Thiele: Communication Synthesis for Reconfigurable Embedded Systems. In *Field-Programmable Logic and Applications*, P. Lysaght, J. Irvine and R.W. Hartenstein (Eds.), pp. 205–214, Berlin: Springer 1999.
20. EN 13306: Maintenance Terminology, 2001.
21. J.C. Geffroy and G. Motet: *Design of Dependable Computing Systems.* Boston: Kluwer 2002.
22. B. Glover, I. Brown and D. West: *Using ISO 9001 to Ensure Reliability in Real-Time Systems.* Des Moines: Nelson Publishing 1995. http://www.evaluationengineering.com/archive/articles/0995Iso.htm
23. M.R. Guthaus, J.S. Ringenberg, T.M. Austin, T. Mudge and R.B. Brown: MiBench: A Free, Commercially Representative Embedded Benchmark Suite. Proc. *4th IEEE Workshop on Workload Characterization*, pp. 3–14, Los Alamitos: IEEE Computer Society Press 2001.
24. R.G. Herrtwich: Echtzeit – Zur Diskussion gestellt. *Informatik Spektrum* 12, 2, 93–96, 1989.
25. C.R. Hofmeister: *Dynamic Reconfiguration of Distributed Applications.* PhD thesis, University of Maryland, Dept. of Computer Science, 1993.
26. G.J. Holzmann: The Power of 10: Rules for Developing Safety-Critical Code. *Computer* 39, 6, 95–97, 2006.
27. B.L. Hutchings and M.J. Wirthlin: Implementation Approaches for Reconfigurable Logic Applications. In *Field-Programmable Logic and Applications*, W. Moore and W. Luk (Eds.), pp. 419–428, Berlin: Springer 1995.
28. IEC 61508: Life-cycle Management of Instrumented Protection Systems. Geneva: International Electrotechnical Commission 1998.
29. IEEE/ANSI Std. 1003.1: Information Technology-(POSIX)-Part 1: System Application: Program Interface (API) [C Language], includes (1003.1a, 1003.1b, and 1003.1c), 1996.
30. ISO/IEC 12207: Systems and Software Engineering — Software Life-cycle Processes, 2008.
31. ISO/IEC 13236: Information Technology — Quality of Service: Framework, 1998.
32. ISO/IEC TR 13243: Information Technology — Quality of Service: Guide to Methods and Mechanisms, 1999.
33. ISO/IEC 15504: Information Technology — Process Assessment, Parts 1–5, 2003–2006.
34. ISO/IEC 17799: Information Technology — Security Techniques: Code of Practice for Information Security Management, 2005.
35. ISO/IEC 27001: Information Security Management Systems — Specification with Guidance for Use, 2005.

36. ISO/IEC 27002: Information Technology — Code of Practice for Information Security Management, 2005.
37. ISO/IEC 9001: Quality Management Systems — Requirements, 2000.
38. ISO/IEC 90003: Software Engineering — Guidelines for the Application of ISO 9001:2000 to Computer Software, 2004.
39. J. Jean, K. Tomko, V. Yavgal, R. Cook and J. Shah: Dynamic Reconfiguration to Support Concurrent Applications. Proc. *IEEE Symp. on FPGAs for Custom Computing Machines*, pp. 302–303, Los Alamitos: IEEE Computer Society Press 1998.
40. Z.T. Kalbarczyk, R.K. Iyer, S. Bagchi and K. Whisnant: Chameleon: A Software Infrastructure for Adaptive Fault Tolerance. *IEEE Transactions on Parallel and Distributed Systems* 10, 6, 560–579, 1999.
41. N.I. Kamenoff and N.H. Weidermann: Hartstone Distributed Benchmark: Requirements and Definitions. Proc. *Real-Time Systems Symp.*, pp. 199–208. Los Alamitos: IEEE Computer Society Press 1991.
42. R.P. Kar and K. Porter: Rhealstone, a Real-time Benchmark Proposal; an Independently Verifiable Metric for Complex Multitaskers. *Dr. Dobb's Journal*, Feb. 1989.
43. R.P. Kar: Implementing the Rhealstone Real-time Benchmark, Where a Proposal's Rubber Meets the Real-time Road. *Dr. Dobb's Journal*, April 1990.
44. G. Kasten, D. Howard and B. Walsh: Rhealstone Recommendations. *Dr. Dobb's Journal*, Sept. 1990.
45. J. van Katwijk, H. Toetenel, A. Sahraoui, E. Anderson and J. Zalewski: Specification and Verification of a Safety Shell with Statecharts and Extended Timed Graphs. In *Computer Safety, Reliability and Security*, F. Koornneef and M. van der Meulen (Eds.), pp. 37–52, LNCS 1943, Berlin: Springer 2000.
46. A.J. Kornecki and J. Zalewski: Software Development for Real-time Safety-critical Applications. Tutorial Notes *29th Annual IEEENASA Software Engineering Workshop*, pp. 1–95, 2005.
47. J. Kramer and J. Magee: Dynamic Configuration for Distributed Systems. *IEEE Transactions on Software Engineering*, 11, 4, 424–436, 1985.
48. C.H.C. Leung: *Quantitative Analysis of Computer Systems*. New York: Wiley 1988.
49. J. Lindström and T. Niklander: Benchmark for Real-time Database System for Telecommunications. *VLDB 2001 Intl. Workshop Databases in Telecommunications II*, LNCS 2209, pp. 88–101, 2001.
50. National Institute of Standards and Technology: Engineering Principles for Information Technology Security. http://csrc.nist.gov/publications/nistpubs/800-27/sp800-27.pdf
51. National Institute of Standards and Technology: Federal Information Processing Standards Publication 197: Announcing the Advanced Encryption Standard. Nov. 2001.
52. F. Nemer, H. Cassé, P. Sainrat, J.P. Bahsoun and M. De Michiel: PapaBench: a Free Real-Time Benchmark. Proc. *Workshop on Worst-Case Execution Time Analysis*, Dresden, 2006.

53. K.M. Obenland: POSIX in Real-Time, 2001. http://www.embedded.com/story/OEG20010312S0073
54. Object Management Group: Real-time CORBA Specification, OMG Document formal/02-08-02 ed., 2002.
55. Object Management Group: UML Profile for Modeling QoS and Fault Tolerant Characteristics and Mechanisms, 2004. http://www.omg.org/docs/ptc/04-06-01.pdf
56. Object Management Group: UML Profile for Schedulability, Performance, and Time Specification 1.1, 2005.
57. Object Management Group: Unified Modeling Language: Superstructure. Version 2.0, 2005.
58. Object Management Group: UML profile for MARTE, Beta 2, June 2008.
59. M.C. Paulk, B. Curtis, M.B. Chrissis and C.V. Weber: Capability Maturity Model for Software, Version 1.1. Software Engineering Institute, CMU/SEI-93-TR-24, 1993.
60. M.C. Paulk, C.V. Weber, S.M. Garcia, M.B. Chrissis and M. Bush: Key Practices of the Capability Maturity Model, Version 1.1. Software Engineering Institute, CMU/SEI-93-TR-25, 1993.
61. W. Perry: *Quality Assurance for Information Systems.* New York: Wiley 1991.
62. W.J. Price: A Benchmark Tutorial. *IEEE Micro* 5, 28–43, 1989.
63. C. Rust, F. Stappert and R. Bernhardi-Grisson: Petri Net Design of Reconfigurable Embedded Real-Time Systems. Proc. *17th IFIP World Computer Congress — Design and Analysis of Distributed Embedded Systems*, pp. 41–50. Dordrecht: Kluwer 2002.
64. B. Schneier: *Applied Cryptography — Protocols, Algorithms, and Source Code in C*, 2nd ed. New York: Wiley 1996.
65. B. Selić, G. Gullekson and P. Ward: *Real-Time Object-Oriented Modeling.* New York: Wiley 1994.
66. B. Selić and J. Rumbaugh: Using UML for Modeling Complex Real-Time Systems. White Paper, Rational Software Corp., 1998.
67. B. Selić: The OSEK/VDX Standard: Operating System and Communication. http://www.embedded.com/2000/0003/0003feat3.htm
68. Seoul National University: SNU Real-Time Benchmarks. http://archi.snu.ac.kr/realtime/benchmark/index.html
69. A.C. Shaw: Communicating Real-Time State Machines. *IEEE Transactions on Software Engineering* 18, 9, 805–816, 1992.
70. A. Southerton, A. Wolfe, Jr., and D. Granz: DOS and UNIX On a Two-way Street, *UNIX World* 9, 8, 48–56, 1992.
71. J.A. Stankovic: Misconceptions About Real-time Computing. *IEEE Computer* 21, 10, 10–19, 1988.
72. T. Swoyer: Balancing Responsiveness, I/O Throughput, and Computation. *Real-Time Magazine* 1, 73–78, 1994.
73. M. Uhle: Leistungstest für Prozeßrechner nach DIN 19 242. *Automatisierungstechnische Praxis* 31, 5, 224–229, 1989.

74. B. Unhelkar: *Verification and Validation of UML 2.0 Models.* Hoboken: Wiley 2005.
75. G.M. Vose and D. Weil: A Benchmark Apologia, Credible Benchmarks are Often the Exception Rather Than the Rule. *Dr. Dobb's Journal*, Feb. 1989.
76. W. Web: Hack This. *Secure Embedded Systems*, July 2004. http://www.edn.com/article/CA434871.html
77. N.H. Weidermann and N.I. Kamenoff: Hartstone Uniprocessor Benchmark: Definitions and Experiments for Real-Time Systems. *Real-Time Systems* 4, 4, 353–383, 1992.
78. W. Wolf: A Decade of Hardware/Software Codesign. *IEEE Computer* 36, 4, 38–43, 2003.
79. W.M. Zuberek: Timed Petri Nets — Definitions, Properties, and Applications. *Microelectronics and Reliability* 31, 4, 627–644, 1991.

Index

Lightning Source UK Ltd.
Milton Keynes UK
16 February 2010

150156UK00005B/163/P